Organic synthesis

Christine L. Willis
Reader in Chemistry, University of Bristol

Martin Wills
Reader in Chemistry, University of Warwick

Series sponsor:  **ZENECA**

ZENECA is a major international company active in four main areas of business: Pharmaceuticals, Agrochemicals and Seeds, Specialty Chemicals, and Biological Products.

ZENECA's skill and innovative ideas in organic chemistry and bioscience create products and services which improve the world's health, nutrition, environment, and quality of life.

ZENECA is committed to the support of education in chemistry and chemical engineering.

OXFORD
UNIVERSITY PRESS

OXFORD

UNIVERSITY PRESS

Great Clarendon Street, Oxford OX2 6DP
Oxford University Press is a department of the University of Oxford.
It furthers the University's objective of excellence in research, scholarship,
and education by publishing worldwide in

Oxford New York

Athens Auckland Bangkok Bogotá Buenos Aires Calcutta
Cape Town Chennai Dar es Salaam Delhi Florence Hong Kong Istanbul
Karachi Kuala Lumpur Madrid Melbourne Mexico City Mumbai
Nairobi Paris São Paolo Singapore Taipei Tokyo Toronto Warsaw

with associated companies in Berlin Ibadan

Oxford is a registered trade mark of Oxford University Press
in the UK and in certain other countries

Published in the United States
by Oxford University Press Inc., New York

© C. L. Willis and M. Wills, 1995

A catalogue record for this book is available from the British Library

Library of Congress Cataloging in Publication Data
(Data available)

Willis, Christine L.
Organic synthesis / Christine L. Willis and Martin Wills.
(Oxford chemistry primers : 31)
Includes bibliographical references and index.
1. Organic compounds–Synthesis.   I. Wills, Martin.   II. Title.
III. Series
QD262.W53 1995
547′.2–dc20   95-14858   CIP

ISBN 0 19 855791 4

Printed in Great Britain by
The Bath Press, Bath

Series Editor's Foreword

Synthesis is the central area of organic chemistry where the chemist's art and imagination are involved as much as his knowledge. As a consequence synthesis is a crucial topic for all students of chemistry but one they often find difficult.

Oxford Chemistry Primers have been designed to provide concise introductions relevant to all students of chemistry and contain only the essential material that would be covered in an 8–10 lecture course. This present primer by Christine Willis and Martin Wills presents the concepts of synthesis in a very logical and student-friendly fashion. This primer will be of interest to apprentice and master chemist alike.

Dr Stephen G. Davies
The Dyson Perrins Laboratory, University of Oxford

Preface

The synthesis of a particular compound from commercially available starting materials is fundamental to nearly all aspects of organic chemistry. There are usually many possible routes for the synthesis of even simple molecules but which is preferred?

The aim of this short text is to give guidelines to enable you to design strategies for the efficient synthesis of a range of molecules including simple mono- and disubstituted compounds and even more complex molecules such as the pyrrolizidine alkaloids. Our approach is based upon retrosynthetic analysis and the importance of bond polarity is emphasised throughout the discussion. In addition we have highlighted some of the mild reagents which are now available to effect chemo-, regio- and stereoselective reactions.

Expertise in synthetic design, as in many aspects of organic chemistry, comes largely with practice. In this book we hope to provide the underlying principles for you to begin to enjoy the challenge of creating efficient routes for the synthesis of organic compounds.

Many thanks to Roger Alder, Alan Armstrong, Stephen Davies, Tina Gleaves, Polly Harrison, Gerry Poulton, Malcolm Sainsbury, John Studley, Tom Simpson, and Heather Tye for their valuable comments on the manuscript.

Bristol   C. L. W.
Bath   M. W.
November 1994

Contents

1 Introduction to synthesis

1.1 The aims and assumptions of this book

In 1965 Professor R. B. Woodward was awarded the Nobel Prize for his contributions over many years to 'The Art of Organic Synthesis', an apt description of a science which requires creativity and ingenuity as much as logical analysis. When considering possible routes for the synthesis of a particular target molecule, one draws on an almost unimaginably vast database of potential transformations and for this reason the subject remains largely a matter of personal opinion and interpretation. In this respect organic synthesis is like the game of chess, in which the rules are simple to learn yet the effective application may take many years to perfect. It is the aim of this book to teach the rules and to provide some hints and guidelines for their application.

In writing this book we have assumed a basic chemical knowledge up to approximately the end of first year degree level although some brief revision of key terms and conventions is provided. If at any stage you find that you are not familiar with any reaction mechanism or concept, we suggest that you refer to one of the texts listed under further reading at the end of each chapter. For a general introduction or revision of key concepts of organic synthesis we recommend the OUP primer *Foundations of Organic Chemistry*, number 9 in this series.

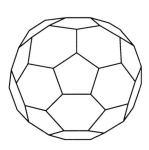

Reserpine, used in the treatment of hypertensive disorders, was synthesised by Woodward in 1956

1.2 The importance of organic synthesis

The synthesis of a particular compound from commercially available starting materials is fundamental to nearly all aspects of organic chemistry. After all, if the physical, chemical, or physiological properties of a compound are to be examined we need a pure sample of it, e.g. the football–shaped molecule buckminsterfullerene (C_{60}). Thousands of organic compounds are naturally occurring and may be isolated from natural sources such as fungi, bacteria and plants but often not in sufficient quantities for a comprehensive study. In addition many fascinating compounds are not naturally occurring and hence these molecules must be prepared to enable their properties to be investigated.

By working steadily through this book we hope that you will learn to draw on your chemical knowledge to design logically efficient syntheses of organic compounds and to assess critically the proposed synthetic route.

One class of natural products which has aroused a great deal of interest from the scientific community is the prostaglandins. These compounds all have a framework based on a five–membered carbocyclic ring bearing two long carbon chains on adjacent atoms, e.g. prostaglandin PGA_2.

Buckminsterfullerene (C_{60})

Prostaglandins have been found in nearly all human tissues and are intimately involved in a host of physiological processes including the induction of labour in childbirth, regulation of the inflammatory response, and the lowering of blood pressure. Initial research on prostaglandins was hampered significantly by the small amounts of material which could be isolated from natural sources, hence many elegant synthetic approaches have been devised.

Prostaglandin PGA$_2$

Consider the problems posed by a synthesis of a molecule such as PGA$_2$: a *trans*–disubstituted cyclopentenone ring must be constructed, the secondary alcohol, ketone, carboxylic acid, and three double bonds must all be placed in their correct positions (regiochemistry) and the geometry of two of the double bonds must be controlled. In addition the stereochemistry of the alcohol and the side chains must be established. To achieve an efficient synthesis of PGA$_2$ therefore presents a significant challenge to the skills of the organic chemist!

Before addressing methods to approach these problems, a brief revision will be given of the importance of bond polarity and the use of the 'curly arrow' notation.

Examples of molecules containing polarised bonds

1.3 Bond polarity

Although the depiction of covalent bonds (e.g. H$_3$C—OH) suggests an even sharing of electrons, this is rarely the case due to the inherent *electronegativity* of atoms. The most electronegative atoms generally lie towards the top right hand corner of the periodic table and are those to which electrons are most strongly attracted. As a result, carbon–heteroatom bonds are polarised and, since most heteroatoms are more electronegative than carbon (i.e. O, N, Br, Cl, I, etc.), a net positive charge resides on carbon. However hydrogen, silicon and metals such as magnesium and lithium are electropositive compared with carbon; the polarity on the C–X bond in these cases is reversed and a net negative charge resides on carbon.

These influences on the electron distribution in σ bonds are known as *inductive effects*.

Bond polarisation effects may be transmitted to remote locations through conjugated π–bonds leading to delocalisation of the charge. For example an α,β–unsaturated carbonyl compound has two sites bearing a partial positive charge as illustrated by the resonance structures shown below (Figure 1.1). By judicious choice of reaction conditions, it is often possible to control the position of attack of a nucleophile to an enone so that only one of these electron deficient sites is attacked. Transmissions of electron distribution through the π–orbitals are known as *mesomeric* effects.

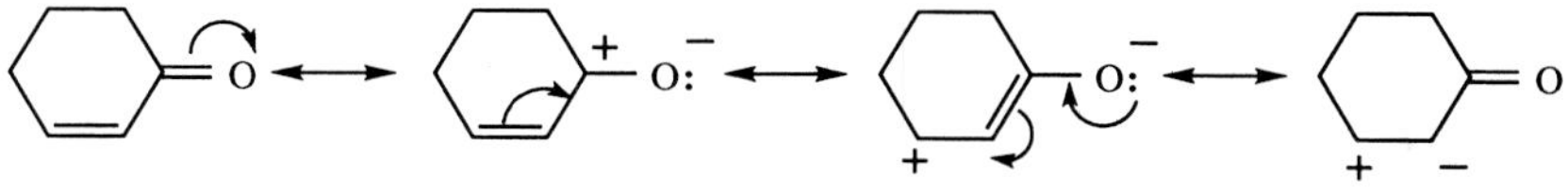

Figure 1.1 Resonance structure of cyclohexenone

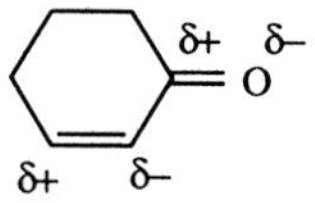

Overall

Although electronegative atoms such as nitrogen and oxygen attract electrons through inductive effects, their lone pair of electrons may be donated through a suitably aligned π–system by mesomeric effects. The resultant patterns of polarisation have an important bearing on the reactivity of organic compounds as illustrated for the case of aniline (aminobenzene), which reacts with electrophiles at the *ortho–* or *para–* positions of the aromatic ring as well as on nitrogen (Figure 1.2).

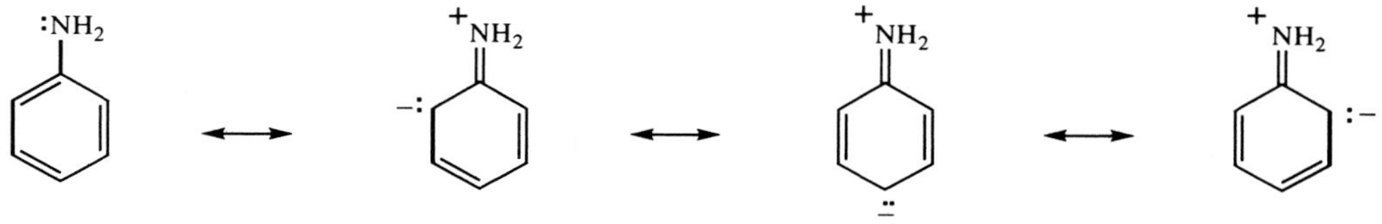

Figure 1.2 Resonance structures of aniline

1.4 'Curly arrow' notation

Arrow pushing to depict mechanism: a brief revision
When one molecule reacts with another, a bond may be formed between an electron deficient atom with a partial (δ+) or full positive charge and an electron rich moiety with a partial (δ–) or full negative charge via electron transfer. 'Arrow pushing' is the convention used by organic chemists to illustrate the *movement of electrons*. Each side of the head of the 'curly arrow' represents one electron. Thus, the single–headed arrows or so–called 'fish hooks' represent the movement of one electron (as in radical reactions, see below) and a double–headed arrow indicates the movement of two electrons. Therefore each double–headed arrow represents:

i) either formation or cleavage of a σ bond e.g. in the S_N2 displacement of an iodide by hydroxide (Figure 1.3) or the protonation of a base (Figure 1.4)

Movement of two electrons

Movement of one electron (fish hook)

Figure 1.3 A typical S_N2 reaction

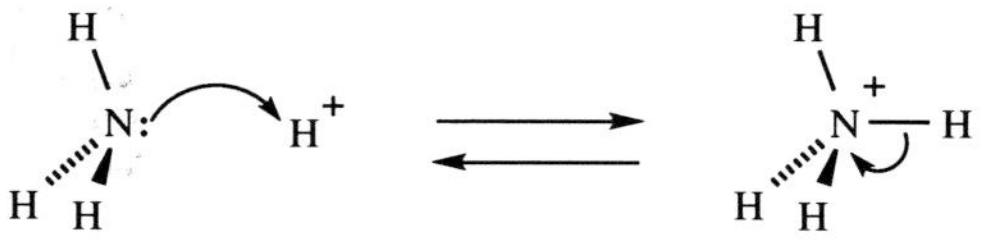

Figure 1.4 Protonation/deprotonation mechanism

ii) the formation or cleavage of a π–bond, e.g. in the hydrolysis of an ester (Figure 1.5). With the exception of protonation/deprotonation, such processes of bond breaking and forming rarely take place in isolation but as part of an extended mechanism.

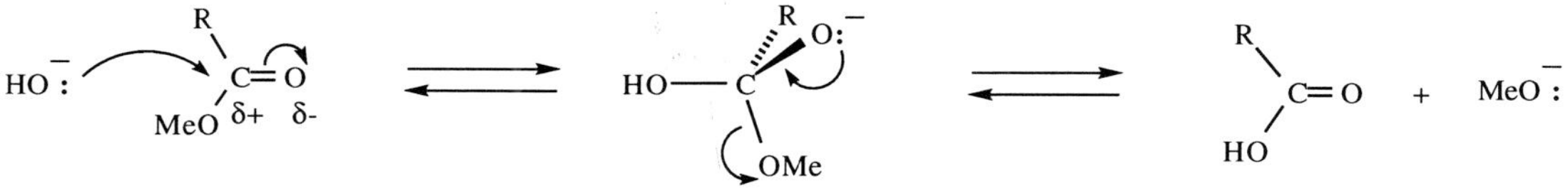

Figure 1.5 Hydrolysis of an ester

Correct

Incorrect

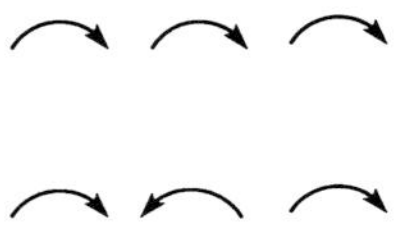

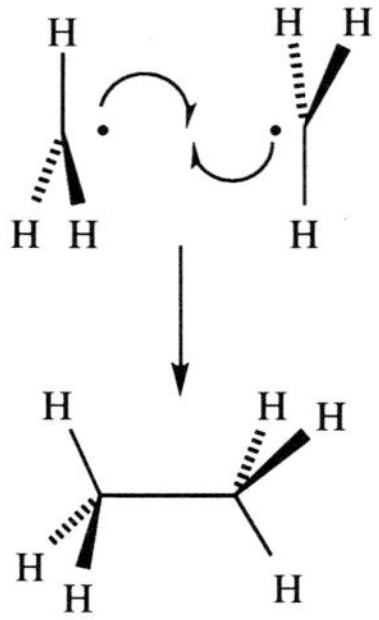

Notice how arrows always 'flow' one after the other. Double-headed arrows in mechanisms must never point away or towards each other. An exception to this is radical reactions, which are described below.

1.5 Free radical reactions

Most organic reactions involve heterolytic cleavage of bonds and reactions between positive or negatively charged or polarised species. However organic compounds can also become involved in radical reactions, an area which is growing in importance in synthesis. In radical reactions the mechanisms are illustrated using single–headed arrows which show the movement of single electrons rather than pairs of electrons. Unlike double–headed arrows, radical arrows *are permitted* to point towards each other, and indeed this is the general case, since each contributes one member of a pair of electrons.

Radical reactions have been used recently for intramolecular cyclisation reactions as illustrated by the synthesis of hirsutene. The key step involves formation of the target molecule from a simple precursor in one impressive radical reaction (Figure 1.6). Homolytic cleavage of the carbon–iodine bond occurs first (initiated by attack by a tributyltin radical), followed by an intramolecular cyclisation and finally reaction with tributyltin hydride gives hirsutene.

Radical reaction mechanism

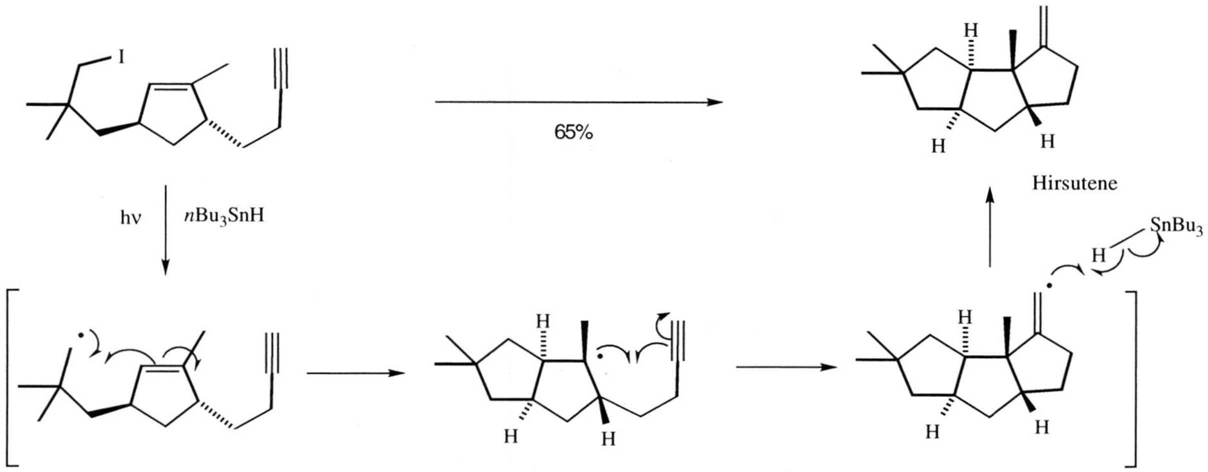

Figure 1.6 The radical cyclisation step in the synthesis of hirsutene

1.6 Conclusions

Further discussions of polarity and its significance in synthetic design are given in Chapters 2 to 4. Many excellent texts are available describing the vast array of chemical reactions which are known, some of which are referenced below. With practice you will not only become familiar with these reactions but also begin to appreciate the evolving patterns of reactivity which give organic chemistry beauty in its simplicity and logic.

Further reading

M. Hornby and J. Peach. *Foundations of Organic Chemistry* (Oxford Chemistry Primer no. 9), Oxford University Press (1993).

K. P. C. Vollhardt and N.E. Schore. *Organic Chemistry* (2nd edn), W. H. Freeman & Co. (1994).

R. O. C. Norman and J. M. Coxon. *Principles of Organic Synthesis* (3rd edn), Blackie (1993).

A. Streitweiser, C. H. Heathcock and E. M. Kosower. *Introduction to Organic Chemistry* (4th edn), Macmillan (1992).

J. McMurry. *Organic Chemistry* (3rd edn), Brooks/Cole (1992).

P. Simpson. *Basic Concepts in Organic Chemistry: A Programmed Learning Approach* (1st edn), Chapman and Hall (1994).

2 Retrosynthetic analysis I: The basic concepts

2.1 Introduction

As stated in Chapter 1, the ability to synthesise a particular compound from commercially available materials is fundamental to nearly all aspects of organic chemistry. When presented with a specific target molecule, a synthetic route must be designed which enables a pure sample of the desired product to be obtained using a convenient and efficient procedure. The following factors need to be taken into consideration:

i) The synthetic design must lead to the construction of the required carbon skeleton with all the substituents and functional groups in the correct positions (regiochemistry) and with the requisite three–dimensional orientations (stereochemistry).

ii) Ideally *the shortest synthetic route* to the product is required. A ten step synthesis with 80% yield at each stage would only give a 10.7% overall yield of the final product.

iii) Each stage of the synthesis should ideally *give only the desired product*. With the armoury of reagents and chemical reactions now at our fingertips, it is often possible to selectively introduce or react one functional group or isomer in an organic molecule when other reactions are possible. However at times mixtures of products are obtained which must be separated by chromatography or distillation.

Chemists have developed a logical approach for the design of routes for the preparation of organic compounds which involves working the synthesis backwards by making strategic carbon–carbon bond cleavages at points where, in the forward direction, bond forming reactions may be achieved. This is known as either the disconnection approach or *retrosynthetic analysis*. The fundamentals of the process of retrosynthetic analysis may be demonstrated by consideration of the synthesis of the secondary alcohol (**1**).

From an examination of (**1**), intuitively you may be able to suggest a possible route for its synthesis and, if you can, there is a high probability that you may be quite unconsciously using your chemical knowledge to achieve an 'informal' retrosynthetic analysis of (**1**). However this 'informal' approach may not be so successful for the synthesis of a more complex molecule, e.g. taxol. In this chapter we have chosen the simple alcohol (**1**) to illustrate the basic principles of retrosynthetic analysis and these fundamental 'guidelines' will be extended to more complex systems in Chapter 3.

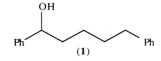

Taxol – a powerful anti–cancer agent

2.2 Synthesis of a target molecule containing one functional group requiring a single disconnection

Retrosynthetic analysis is the process of 'breaking down' a target molecule into readily available starting materials by means of imaginary breaking of bonds (*disconnections*) and by the conversion of one functional group to another by efficient chemical reactions (*functional group interconversions*). What is meant by 'readily available or simple starting materials' is often not obvious when you first begin to think about synthesis. As a guide, readily available starting materials contain five carbon atoms or less (apart from units such as aromatic rings) and only one or two functional groups. However there are many exceptions to this general rule and when designing routes for the synthesis of organic compounds, it is valuable to refer to catalogues of commercially available materials.

In this case we will use retrosynthetic analysis to break down the alcohol (**1**) (our target molecule, T.M.), into readily available starting materials and then use this analysis to devise an appropriate 'forward' synthetic route to the T.M. There is often more than one 'correct' analysis. For example, even for a relatively simple molecule such as (**1**) we can easily think of at least six different ways to work backwards to break the molecule down into readily available starting materials. These six methods will now be described to highlight the principles of retrosynthetic analysis and then the relative merits of each route will be assessed.

Retrosynthetic analysis 1 of (1)

First, imagine breaking (disconnecting) the carbon–carbon bond of (**1**) shown in Figure 2.1 and putting a positive charge on one end of the cleaved bond and a negative charge on the other fragment.

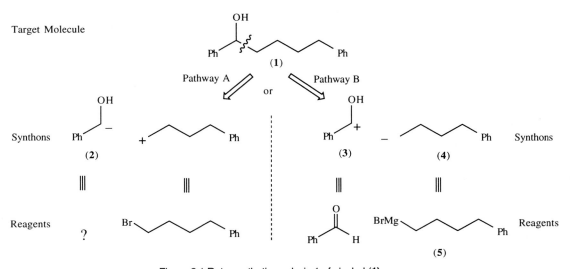

Figure 2.1 Retrosynthetic analysis 1 of alcohol (**1**)

The disconnection is represented by a wavy line through the bond to be broken and the retrosynthetic arrow ⇒ represents going from the target molecule 'backwards' to the pair of charged fragments (known as *synthons*). In theory this disconnection may lead to two pairs of idealised fragments or synthons from which the alcohol can be prepared. If at first you are not sure which charge to put on which fragment, try both to see which of the two alternatives relates to recognisable functional groups in readily available starting materials.

In this case, since oxygen is more electronegative than carbon, it is *not* easy to think of a simple reagent which will give the polarity on the carbon atom of synthon (**2**) resulting from pathway A. However considering pathway B, benzaldehyde and the Grignard reagent (**5**) are readily available reagents equivalent to the synthons (**3**) and (**4**). Therefore from this analysis it is apparent that a straightforward synthesis of (**1**) would be the reaction of the Grignard reagent and benzaldehyde as shown in Figure 2.2.

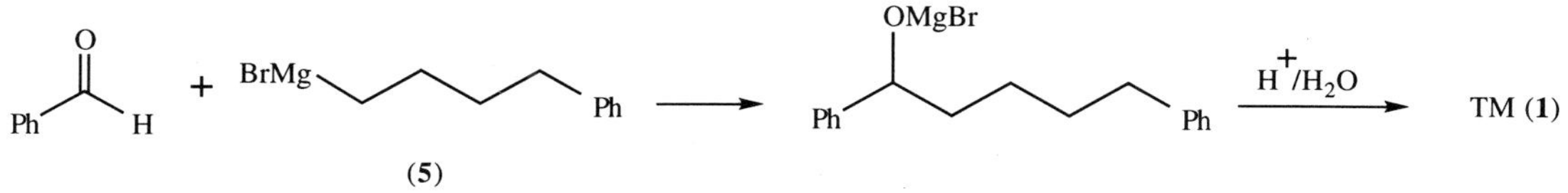

Figure 2.2 Synthesis 1 of (**1**)

Retrosynthetic analysis and synthesis 2 of (1)

Another equally possible retrosynthetic analysis of (**1**) involves disconnecting the carbon–carbon bond shown in Figure 2.3.

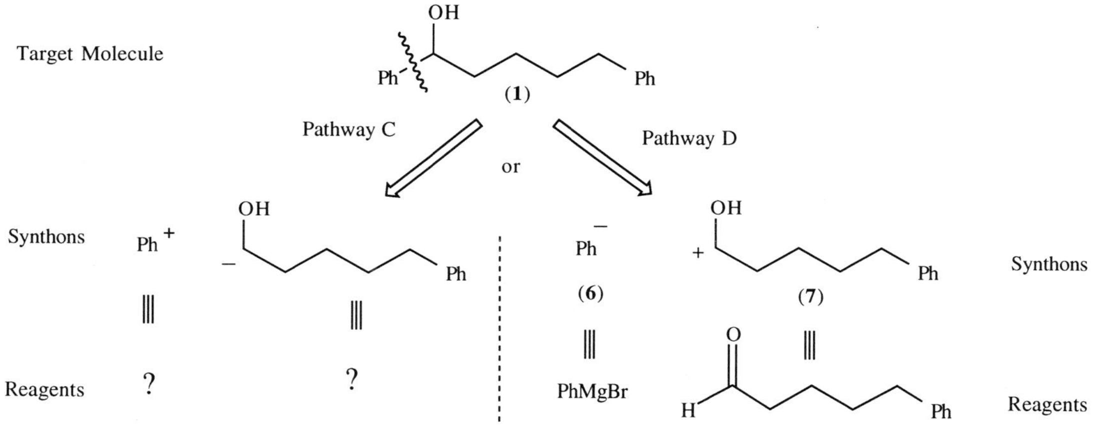

Figure 2.3 Retrosynthetic analysis 2 of alcohol (**1**)

Again this leads to two possible pairs of ionised fragments. Only pathway D gives a pair of synthons (**6**) and (**7**), for which there are obvious simple equivalent reagents, i.e. a Grignard reagent and an aldehyde. The synthetic route derived from this analysis is shown in Figure 2.4.

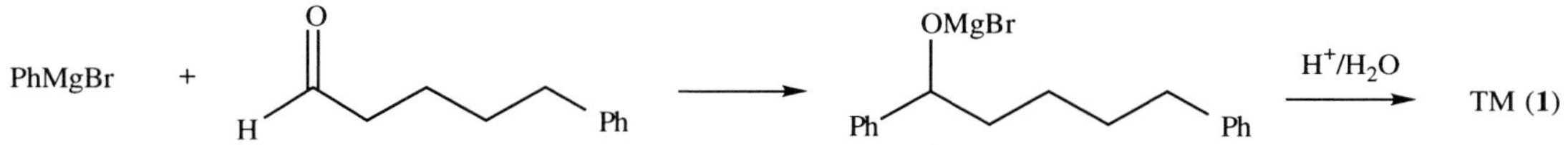

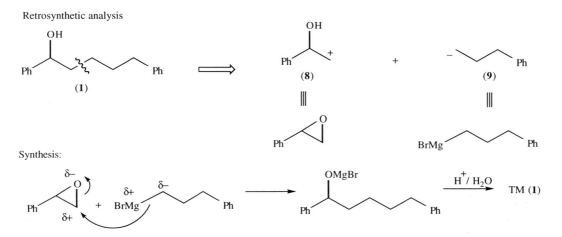

Figure 2.4 Synthesis 2 of alcohol (**1**)

Retrosynthetic analysis and synthesis 3 of (1)

As a rule, an unsuitable pair of synthons which may arise from a disconnection is not shown in the final written version of the retrosynthetic analysis although it may have been taken into consideration when choosing the eventually favoured route for the synthesis of the target molecule. Therefore a third retrosynthetic analysis of (**1**) may be represented as shown in Figure 2.5. The favoured pair of synthons (**8**) and (**9**) are equivalent to the epoxide and the Grignard reagent which leads to the simple synthetic route shown in the same figure.

Figure 2.5 Retrosynthetic analysis and synthesis 3 of (**1**)

Retrosynthetic analysis and synthesis 4 of (1)

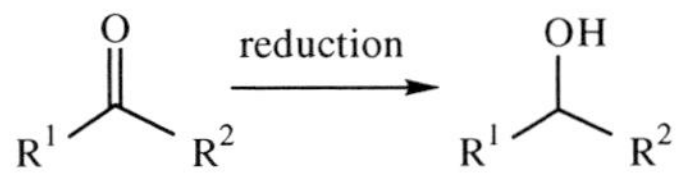

In each of the three synthetic routes described above, the alcohol (**1**) has simply been prepared by the reaction of a Grignard reagent with either an aldehyde (Methods 1 and 2) or an epoxide (Method 3). A different approach to the synthesis of (**1**) derives from the knowledge that ketones may be cleanly reduced to give secondary alcohols with reagents such as sodium borohydride or lithium aluminium hydride.

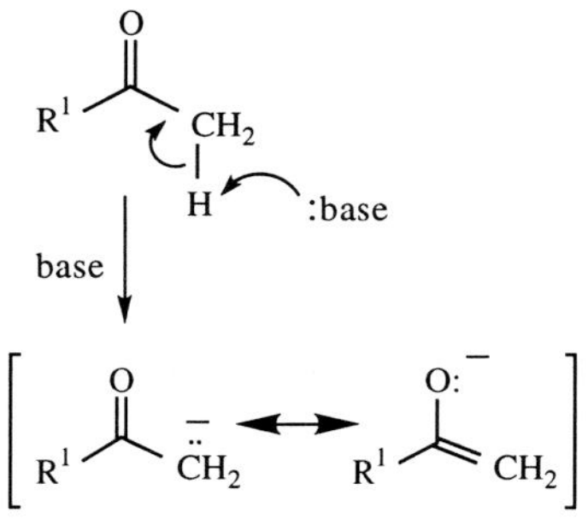

Functional group interconversion (FGI) is the term used in retrosynthetic analysis to describe the process of converting one functional group into another, for example by oxidation or reduction. It is represented by the symbol $\Rightarrow$ with 'FGI' above it.

Therefore if the alcohol (**1**) is converted into a ketone as the first stage of the analysis prior to cleavage of the carbon–carbon bond shown in Figure 2.6, then the pair of synthons (**10**) and (**11**) is simply equivalent to the addition of the enolate of acetophenone to the halide giving the synthesis shown below (Figure 2.6).

Remember that protons α to a carbonyl group are acidic and may be abstracted with base to give an enolate.

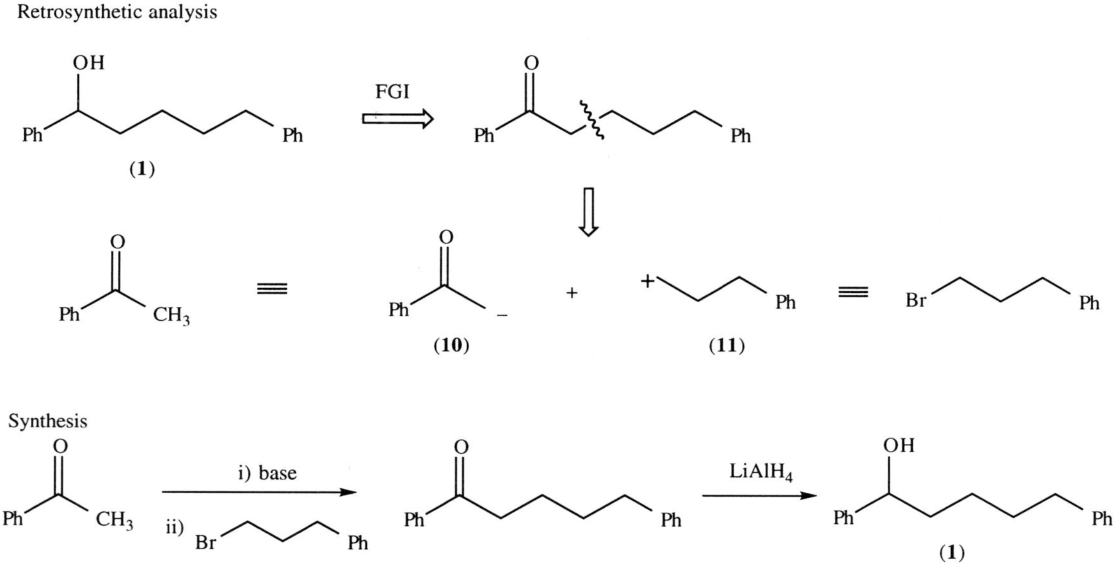

Retrosynthetic analysis

Synthesis

Figure 2.6 Retrosynthetic analysis and synthesis 4 of (**1**)

Retrosynthetic analysis and synthesis 5 of (1)

A further retrosynthetic analysis of (**1**) (Figure 2.7) again involves a functional group interconversion of the alcohol to the ketone prior to cleavage of a carbon–carbon bond. This gives the synthon (**12**) with the positive charge β to the carbonyl group (for which we can use the enone (**14**) as the synthetic equivalent), and the synthon (**13**) for which an appropriate carbon nucleophile such as a dialkylcuprate or a copper catalysed Grignard reagent (see Chapter 6) may be used (Figure 2.7).

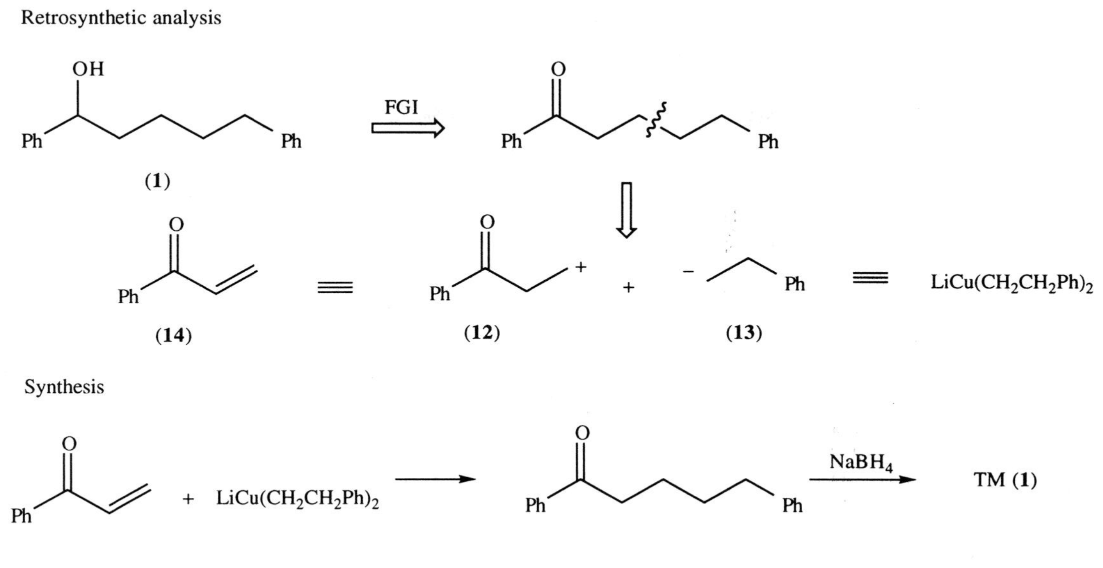

Retrosynthetic analysis

Figure 2.7 Retrosynthetic analysis and synthesis 5 of alcohol (1)

Retrosynthetic analysis and synthesis 6 of (1)

The final retrosynthetic analysis of (1) to be considered again requires an initial functional group interconversion of the alcohol to a ketone followed by a second FGI to form the α,β–unsaturated ketone (15). Cleavage of the carbon–carbon bond shown in Figure 2.8 leads to a synthesis in which addition of lithium diphenylcuprate to the dienone (16) furnishes the required carbon skeleton. Reduction of the enone (15) with sodium borohydride in the presence of copper (I) ions leads to 1,4–reduction (see Chapter 6) to give the requisite secondary alcohol (1) (Figure 2.8).

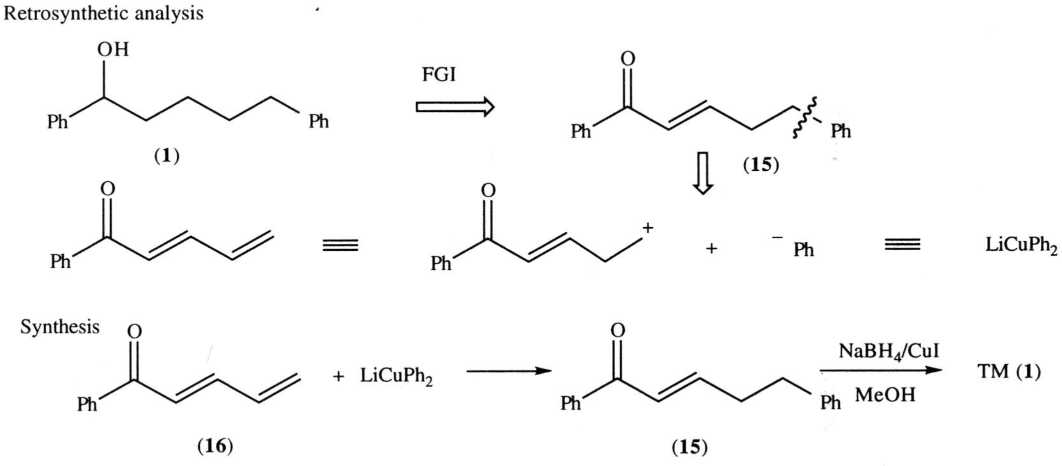

Retrosynthetic analysis

Figure 2.8 Retrosynthetic analysis and synthesis 6 of alcohol (1)

The merits of the six synthetic routes to (1)

The six methods for the synthesis of (1) that have now been considered may be summarised as shown in Figure 2.9.

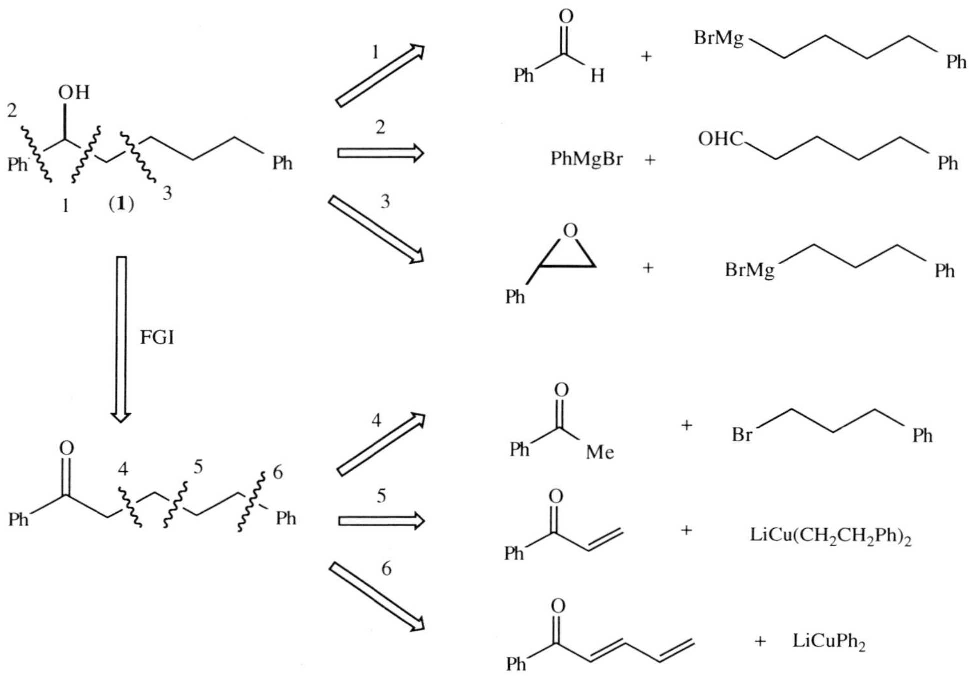

Figure 2.9 Summary of disconnections of alcohol (1)

Which is the best method?

The route which is eventually chosen for the synthesis of (1) may depend on a variety of factors including the cost and availability of the reagents.

As a general rule, disconnections nearer the centre of a molecule usually lead to the greatest simplification as will be seen in Chapters 3 and 4; therefore methods 1, 3, and 4 may be preferred.

In addition, the number of synthetic steps should ideally be kept to a minimum *unless* there is an advantage in using an FGI to facilitate a high yielding carbon–carbon bond forming reaction. By this criterion methods 1, 2, and 3 each require only a one–pot synthesis of (1) and would therefore be on our initial short list for investigation. However lithium dialkylcuprate additions to enones have proved to be excellent methods for the formation of carbon–carbon bonds; therefore method 5 should be given serious consideration.

In retrosynthetic analysis there is often more than one 'correct' answer and the route eventually chosen may come down simply to personal choice. The advantages of retrosynthetic analysis may not be apparent to you from the example of the synthesis of the simple alcohol (**1**), but its value should become more obvious when we consider more complex target molecules.

2.3 Synthetic equivalents to common synthons

In all the cases described above, the importance of the concept of polarity of bonds within functional groups (as outlined in Chapter 1) should be apparent, e.g. the epoxide with a δ– on oxygen and a δ+ on carbon enabling attack by the electron rich carbon atom of the Grignard reagent to form the required carbon–carbon bond. It is worth familiarising yourself with some of the synthetic equivalents which correspond to common synthons. These are given in Table 2.1 and may be deduced directly from a consideration of bond polarity and resonance effects.

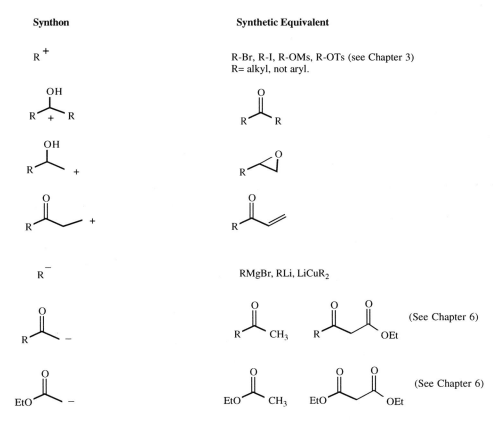

Table 2.1 Synthons and their synthetic equivalents

2.4 Practice examples

Expertise in synthetic design, as in many aspects of organic chemistry, comes largely with practice and hence we recommend that you attempt the following examples before moving on to Chapter 3. In each case use retrosynthetic analysis to design a synthesis for each of the following molecules using the suggested starting materials and any other necessary reagents.

The terms introduced in this section, retrosynthetic analysis, target molecule, disconnection, synthon, functional group interconversion and synthetic equivalent are defined in the Glossary.

Further reading

J. Fuhrhop and G. Penzlin. *Organic Synthesis: Concepts, Methods and Starting Materials*, Verlag Chemie (1983).
S. Warren. *Designing Organic Syntheses: A Programmed Introduction to the Synthon Approach*, Wiley (1978).
S. Warren. *Organic Synthesis: The Disconnection Approach*, Wiley (1982).
E. J. Corey and X.–M. Cheng. *The Logic of Chemical Synthesis*, Wiley Interscience (1989).

3 Retrosynthetic analysis II:
Latent polarity and FGIs

3.1 Introduction

From the discussion so far it should be evident that the carbonyl group plays a key role in organic synthesis. Indeed carbonyl compounds were used in five out of the six syntheses of the alcohol (**1**) suggested in Chapter 2.

There are two basic reactions which the carbonyl group may undergo:

i) addition of a nucleophile to the carbonyl group, either under basic conditions:

or under acidic conditions:

ii) deprotonation of the carbon α to the carbonyl group followed by reaction of the resultant enolate with an electrophile:

Alternatively a ketone may react with an electrophile under acidic conditions *via* an enol.

Unsaturated carbonyl compounds undergo similar reactions to saturated carbonyl compounds as shown in Figure 3.1. Further details of the use of selective reagents to control this reactivity are given in Chapter 6. Notice how the reactivity of the enone (**2**) towards electrophiles and nucleophiles alternates as you move down the chain from the carbonyl group. Indeed the pattern of alternating electrophilic and nucleophilic sites may be continued down an unsaturated hydrocarbon chain *provided that double bonds are present in conjugation with the carbonyl group* (i.e. with an alternating pattern of double and single bonds).

i) Addition of a nucleophile

ii) Deprotonation followed by reaction of resulting enolate with an electrophile at the α or γ positions

or

Figure 3.1 Reactions of unsaturated carbonyl compounds

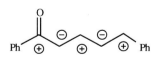

Note that in this text latent polarities will be represented as the circled forms ⊕ and ⊖ to distinguish them from full charges. However please note that this should not be regarded as a widely accepted notation.

In the previous chapter the basic concepts of retrosynthetic analysis were introduced and the importance of polarity highlighted. The reactivity of unsaturated carbonyl compounds introduces the use of a further pattern of bond polarity which is emerging in our logical approach for the design of efficient syntheses of target molecules. The best disconnections (all but disconnection 3) of target alcohol (**1**) shown in Chapter 2 (summarised in Figure 2.9) led to synthons which had an alternating pattern of charges as we move away from the hydroxyl or carbonyl group, starting with a positive charge on the carbon adjacent to oxygen. Writing this alternating pattern of *imaginary* charges or 'latent polarities' on a target molecule can greatly assist the identification of potential synthons since breaking a bond in the retrosynthetic direction immediately gives viable synthons for which corresponding reagents are available.

Many functional groups commonly found in organic molecules exhibit a similar pattern of latent polarities to the carbonyl and hydroxyl group. For example alkyl halides (which may be prepared from alcohols), imines and amines have the same pattern of latent polarities as a carbonyl group (Figure 3.2). Note that in all cases, the polarities are arranged such that a positive charge is initially placed on the carbon adjacent to the heteroatom, whether linked through a single or double bond. Exceptions to these patterns will be highlighted as appropriate in the following discussion.

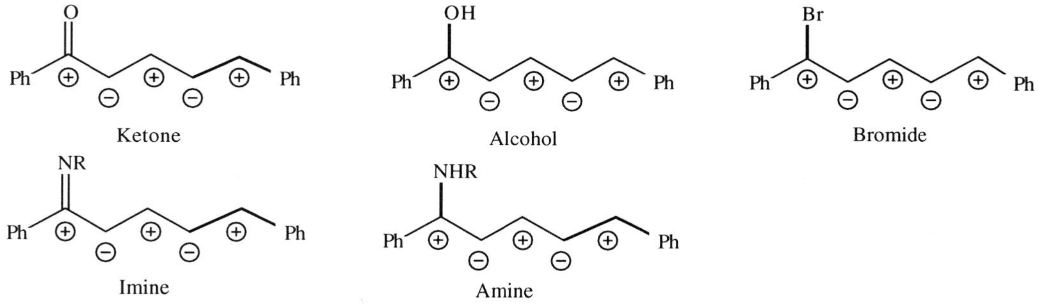

Figure 3.2 Latent polarity is the imaginary pattern of alternating positive and negative charges used to assist in the identification of disconnections and synthons

3.2 Target molecules with two functional groups

If a target molecule contains more than one substituent or functional group, it is essential that the synthesis is designed to take account of the final position of all of these groups. The concept of latent polarity often enables us to identify the best position to make a disconnection within more complex molecules.

1,3–Disubstituted compounds

Consider the synthesis of the β–hydroxy ketone (**3**). If the latent polarities arising from each of the functional groups are added, we see that they overlap (Figure 3.3). This coincident relationship of overlapping latent polarities is known as a *consonant* pattern and when this pattern is present within a molecule, a simple synthesis may often be achieved.

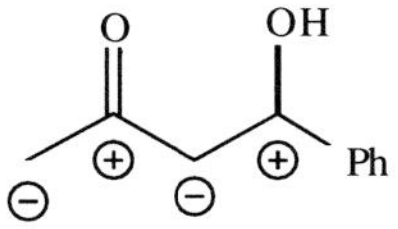

Examining the retrosynthetic analysis of (**3**), it is apparent that one of the best disconnections leads to the synthons (**4**) and (**5**) corresponding to benzaldehyde and the enolate of acetone (Figure 3.4). Under appropriate conditions this molecule may dehydrate to give the unsaturated ketone (Figure 3.5).

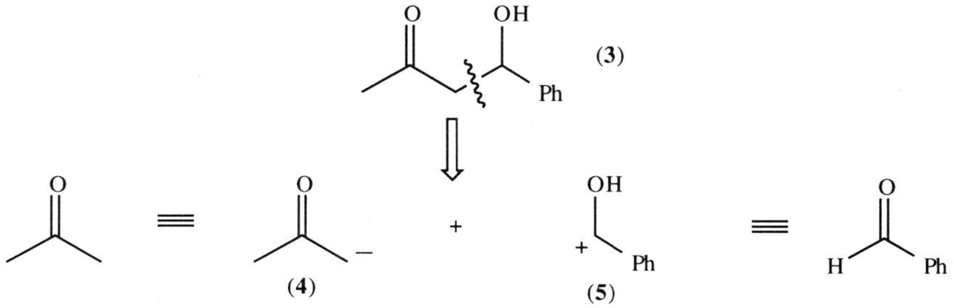

Figure 3.3 Latent polarities of the carbonyl and hydroxyl groups in (**3**) may be superimposed

Figure 3.4 Retrosynthetic analysis of (**3**)

Figure 3.5 Synthesis of (3)

This simple synthesis neatly puts both functional groups in the β-hydroxy ketone (3) in their correct position. A similar approach may be used for the synthesis of 1,3–dicarbonyl compounds and 1,3–diols, since these compounds may be simply prepared from the β–hydroxy ketone (3) by functional group interconversions (Figure 3.6).

Figure 3.6   Reactions of (3)

1,5-Disubstituted   compounds

As a general rule *a disconnection nearer the centre of a molecule is usually favoured so that maximum simplification is achieved.* As a logical extension to the above example, the retrosynthetic analysis of the 1,5–dicarbonyl compound (6) reveals a similar matching or consonant pattern of latent polarities (Figure 3.7). In this case the best synthon pair is (7) and (8) corresponding to the enone (9) and enolate derived from acetone as shown in Figure 3.8. The synthesis then is a conjugate addition (or Michael addition) of the enolate to the enone.

Simplicity is the key. Carbonyl reactivity alone is enough to control the location of functional groups in the target molecule (6). Not only may compounds containing 1,3– or 1,5– dioxygenated functions be disconnected in this way, any system containing a 1,*n*–dioxygenated function where *n* is an odd number follows this pattern, provided the reagents for its synthesis are unsaturated to enable transmission of polarity. The extensive use of the carbonyl group throughout this sequence underlines its crucial significance in synthesis.

(6)

Figure 3.7 Latent polarities of the carbonyl groups in (6) may be superimposed

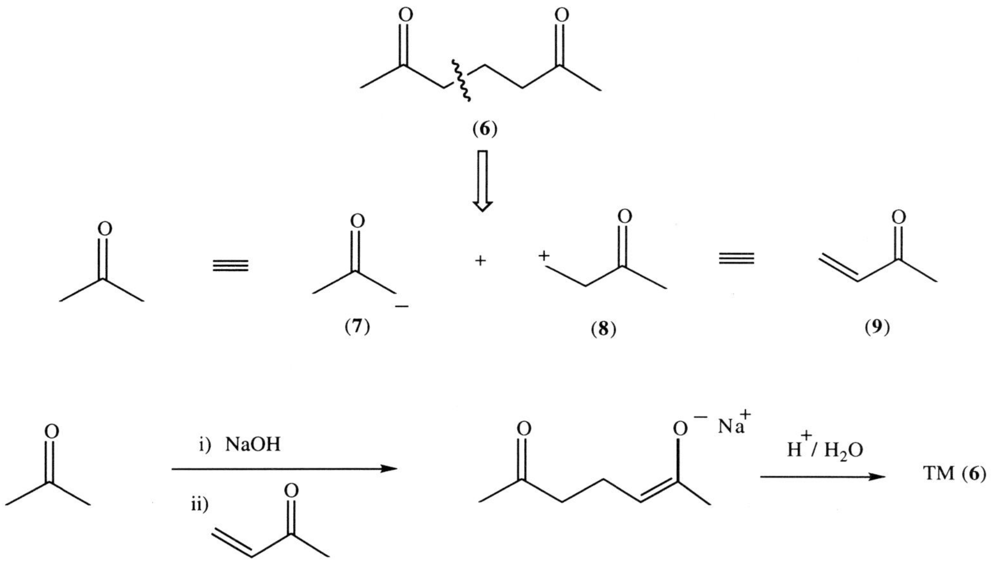

Figure 3.8 Retrosynthetic analysis and synthesis of (6)

3.3 1,4–Dicarbonyl compounds and umpolung (reversal of polarity)

The process of identifying suitable disconnections becomes rather more complex when groups in a T.M. form patterns of latent polarity which do not match up. Consider the synthesis of 2,5–hexadione (**10**). If the patterns of latent charges are added to (**10**) it becomes apparent that they are not superimposable (Figure 3.9). This relationship is termed *dissonant*. In this case we cannot disconnect the system to a pair of precursors which can be linked simply under the control of the reactivity of the carbonyl group.

Retrosynthetic analysis of (**10**) (Figure 3.10) may lead to synthons (**11**) and (**12**); it is clear that whilst the former corresponds to the enolate of acetone, the charge in (**12**) does not correspond to the polarity normally associated with a carbonyl group or its derivatives. We therefore require a reagent equivalent to synthon (**12**). The German word *umpolung* is used to describe situations of this sort in which a synthon of opposite polarity to that normally associated with the required functional group must be used. A great deal of research has been targeted towards devising solutions to this type of disconnection problem and some commonly used methods are outlined below.

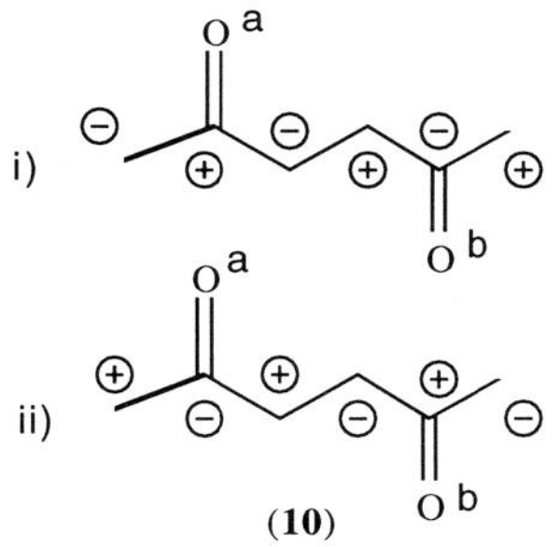

Figure 3.9 Pattern of latent polarities in (**10**) due to:
i) carbonyl group a
ii) carbonyl group b

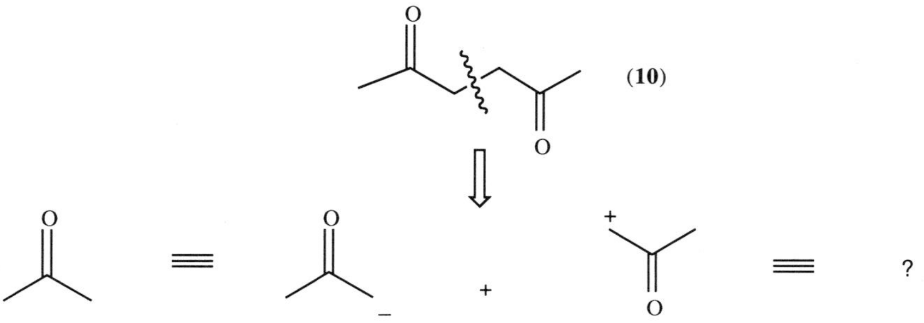

(10)

(11)   (12)   ?

Figure 3.10 Retrosynthetic analysis of (10)

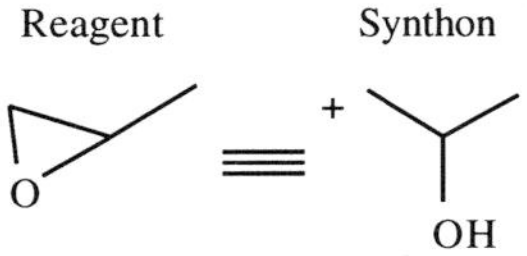

Reagent   **Synthon**

a) Epoxides. We have already seen an example of the utility of the nucleophilic ring opening of an epoxide in the preparation of the alcohol (1) (Chapter 2, Figure 2.5). Reaction of epoxides with nucleophiles is a relatively facile process due to the relief of strain energy achieved upon opening the oxirane ring. If the nucleophile is an enolate, the product will contain a 1,4–dioxygenated pattern. Subsequent adjustment of the oxidation level of the 1,4–dioxygenated compound may then be used (in the case of (10) an oxidation is required, Figure 3.11). Further details of the value of epoxides in synthesis are given later.

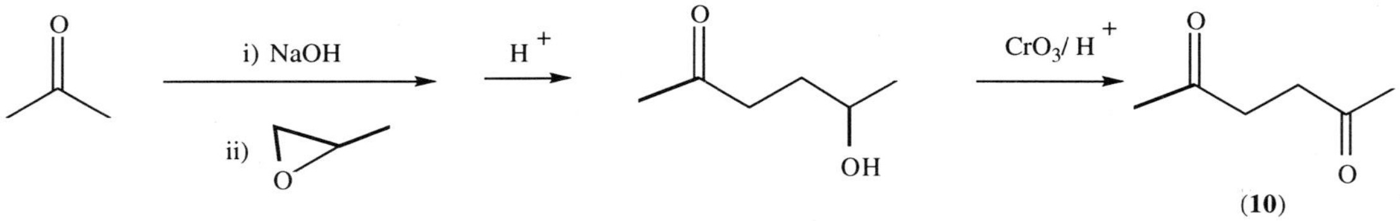

i) NaOH   H$^+$ CrO$_3$/ H$^+$

ii)   (10)

Figure 3.11 Use of an epoxide in the synthesis of a 1,4–dioxygenated compound

Reagent   **Synthon**

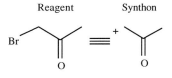

b) α–Haloketones or esters. Perhaps the simplest solution to the problem of mismatched latent polarities is to use reagents with strategically placed heteroatoms. The leaving group ability of the bromide overrides the inherent latent polarity pattern imposed by the carbonyl group (Figure 3.12). α-Bromoketones or esters are generally prepared by the acid catalysed bromination of the parent ketone or ester.

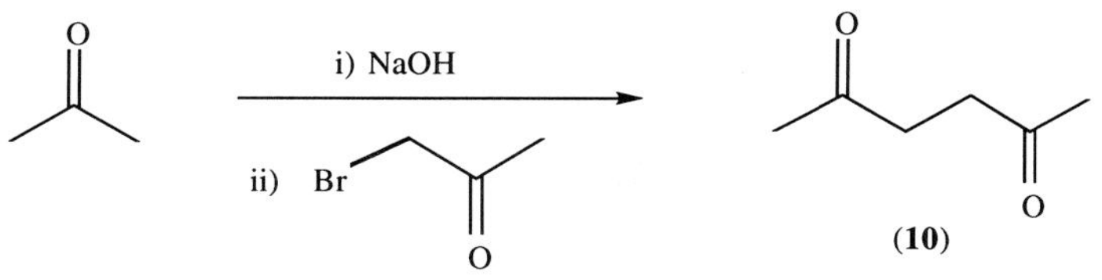

Figure 3.12 Use of an α-bromoketone in the synthesis of a 1,4–diketone

A word of caution: α-halo carbonyl compounds may be readily deprotonated at the position adjacent to the halide and the resultant enolate may then attack another carbonyl group giving an epoxy-ketone as shown in Figure 3.13. This reaction is known as the Darzens condensation. Although this is a valuable reaction, it does not give the product required in this case. A way to circumvent the problem is to use an enamine instead of an enolate (for further details see recommended texts). The Favorskii rearrangement may also be a competing side reaction (see OUP primer *Polar Rearrangements*, number 5 in this series).

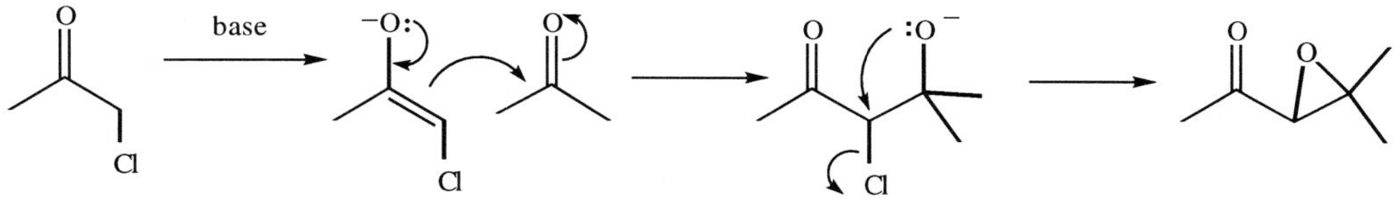

Figure 3.13   The Darzens condensation

c) **1,3-Dithianes**.   Reaction of a ketone or an aldehyde with 1,3–propanedithiol in the presence of an acid catalyst under dehydrating conditions gives a 1,3–dithiacyclohexane (1,3–dithiane). The hydrogen atoms on the carbon atom between the two sulphur atoms are relatively acidic (pK_a 31) so that reaction with a strong base such as butyllithium results in the formation of the corresponding anion.   The anion will react with a variety of electrophiles including carbonyl compounds, epoxides and primary and secondary halides. Finally the thioacetal function may be hydrolysed in good yield by mercuric salts to give the corresponding carbonyl compound (Figure 3.14).   Thus the electrophilic carbonyl compound has undergone reverse polarisation in the formation of the thioacetal (to become the nucleophile). The 1,3–dithiane is known as an *acyl anion* equivalent.

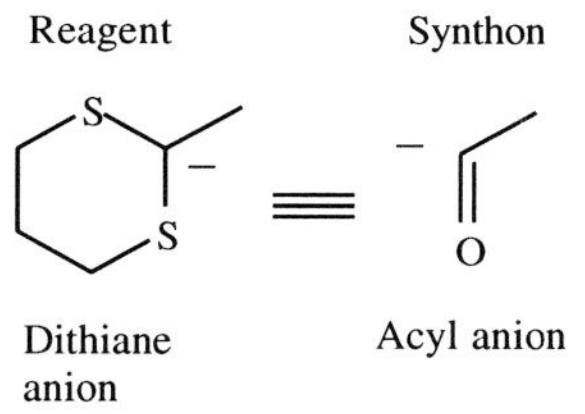

Reagent	Synthon

Dithiane anion / Acyl anion

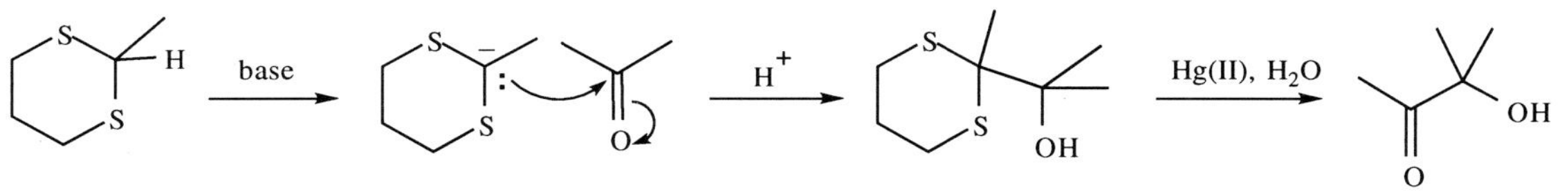

Figure 3.14 1,3-Dithianes as acyl anion equivalents

This procedure has been put to good use in the total synthesis of (**11**), the sex attractant of the bark beetle (Figure 3.15).

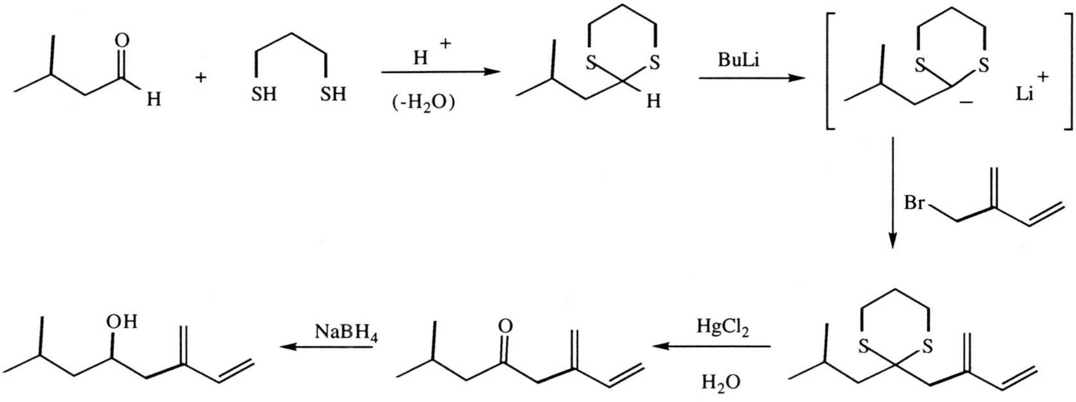

Figure 3.15 Synthesis of bark beetle pheromone (**11**)

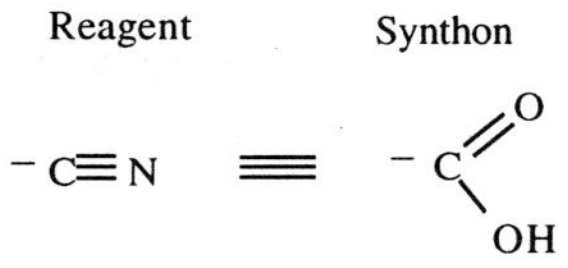

d) The addition of cyanide. Hydrogen cyanide (care: volatile and highly toxic) adds to ketones and aldehydes to give adducts known as cyanohydrins. The cyanide function may then be further modified, e.g. hydrolysis of the cyanohydrin gives the 2–hydroxy–carboxylic acid. In this reaction sequence the cyanide ion has effectively acted as the synthetic equivalent of the synthon (**12**), in which a negative charge is located on the carbon atom directly attached to oxygen (Figure 3.16). Further reactions of the cyanide (or nitrile) group are given later in this chapter.

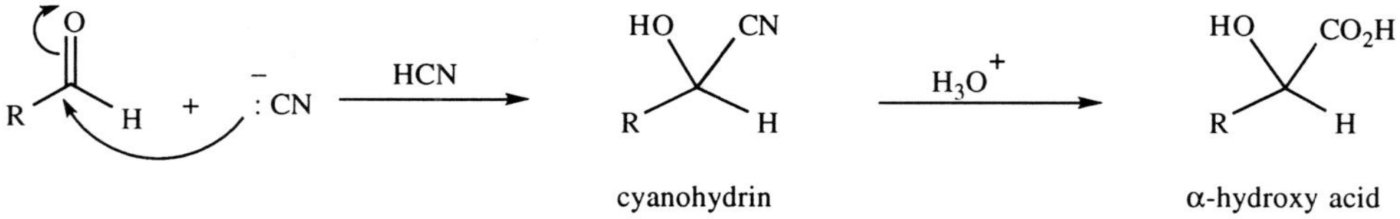

Figure 3.16 Use of the cyanide anion as anionic carboxylic acid equivalent

Catalytic sodium cyanide has been used to good effect in the *benzoin condensation* in which α–hydroxy ketones are prepared by the dimerisation of aromatic aldehydes. The reaction proceeds essentially by the reversal of the polarity of one of the carbonyl groups by the cyanide (Figure 3.17). The cyanide ion may be used *only* for coupling non-enolisable aldehydes, usually derivatives of benzaldehyde. In this case the reversal of polarity is in the form of *reactive intermediates*. Compare this with the case of the 1,3–dithianes which may be isolated.

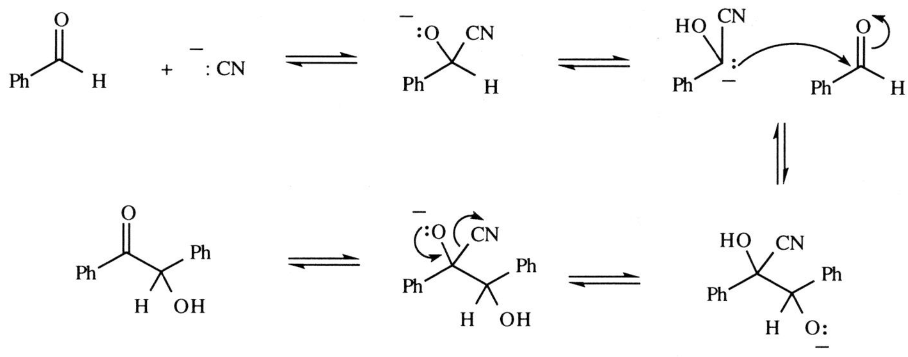

Figure 3.17 The benzoin condensation

e) The nitro group. Deprotonation at the position adjacent to a nitro group may be achieved with a number of bases (n.b. nitromethane has pK_a 10.2). The resultant anion is nucleophilic and will undergo typical nucleophilic reactions, e.g. addition to an aldehyde, the Henry reaction (Figure 3.18).

Hydrolysis of the nitro group gives an aldehyde. This transformation is known as the Nef reaction. Overall the nitromethane anion is equivalent to a *formyl anion*.

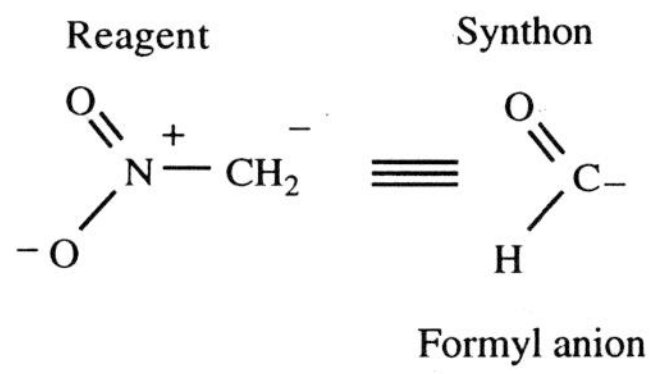

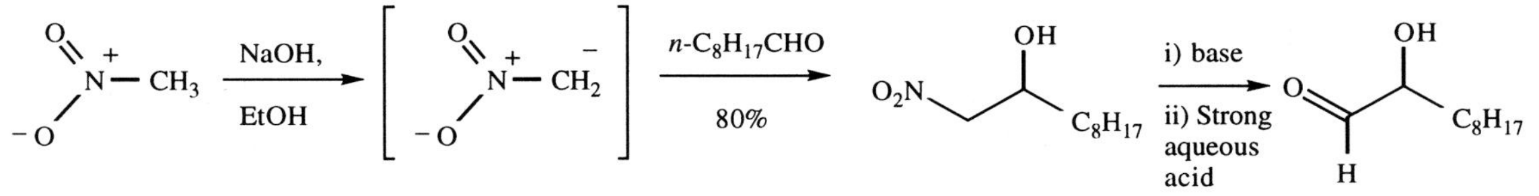

Figure 3.18 The Henry reaction followed by the Nef reaction

f) Alkynes. Deprotonation of an alkyne (e.g. ethyne, pK_a 25) with a strong base results in formation of the anion (**13**). The anion reacts with a range of electrophiles (see Chapter 6). Subsequent hydrolysis of the triple bond gives a ketone (Figure 3.19). This is a further example of an acyl anion equivalent since the acetylene anion is effectively equivalent to the synthon (**14**).

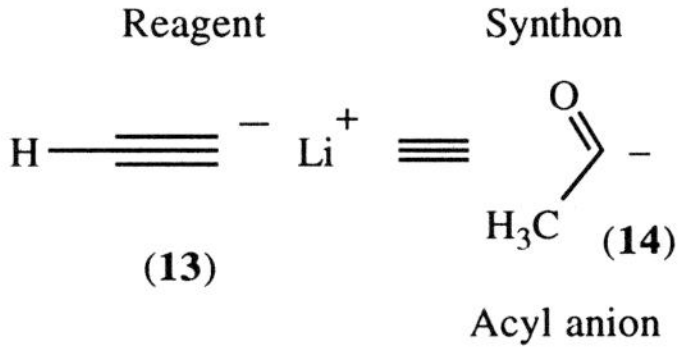

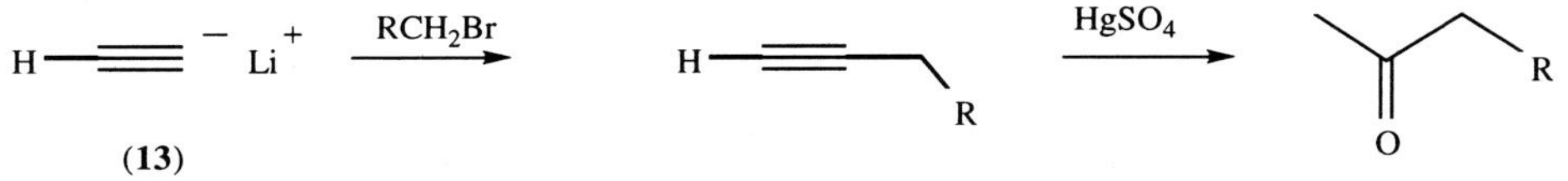

Figure 3.19 Alkyne anion as an acyl anion equivalent

3.4 Practice examples

Using retrosynthetic analysis, suggest syntheses of the following compounds using starting materials containing no more than seven carbon atoms.

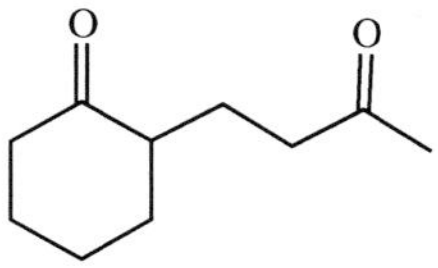

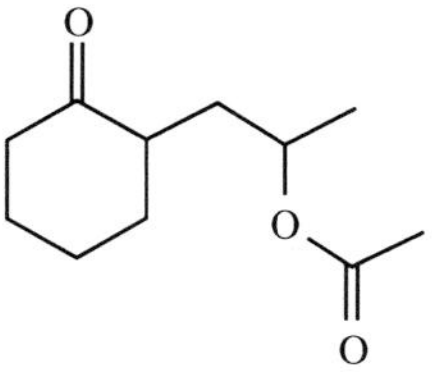

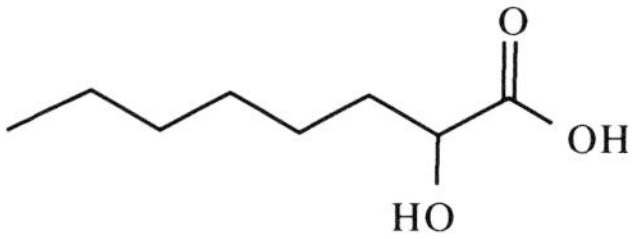

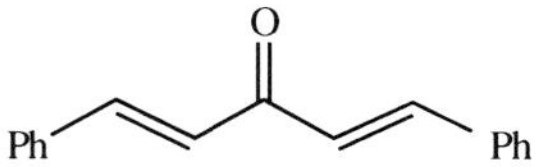

3.5 Synthesis of cyclic molecules

The retrosynthetic analysis and hence the synthesis of cyclic systems follows essentially the same rules as for acyclic systems. However a larger choice of possible disconnections generally exists if more than one functional group is to be incorporated into the cyclic product. Most of the reactions described above may be used in an *intramolecular* sense to build rings. Several factors have to be considered:

i) *The probability of the two reactive species meeting to enable an intramolecular reaction to proceed.* There is a higher probability of two reactive species colliding when separated by a shorter rather than a longer chain. Indeed if the intramolecular cyclisation reaction would lead to a ring larger than six–membered, intermolecular reactions may compete with the cyclisation process leading to the formation of dimers, trimers and other polymers. Reaction of the diester (**15**) with sodium ethoxide gives the cyclopentanone (**16**) (Dieckmann condensation) in 80% yield whereas cyclisation of (**17**) to form the cyclohexanone (**18**) proceeds in only 54% yield (Figure 3.20).

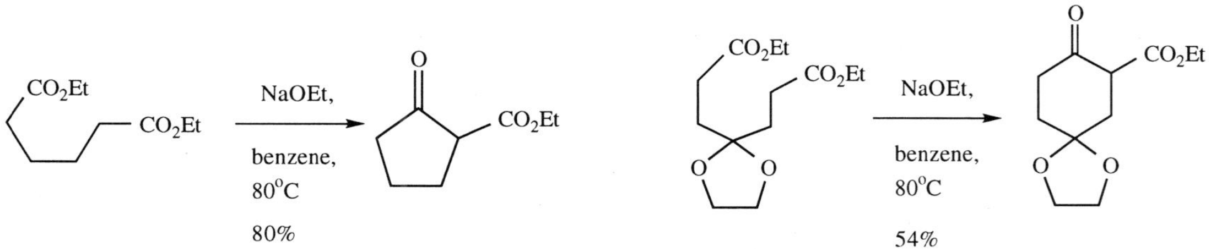

Figure 3.20 The use of the Dieckmann condensation for the synthesis of five and six membered rings

Large carbocyclic rings may be cleanly prepared from diesters by the acyloin reaction in which the ester functions form reactive radical anions on the metal (Figure 3.21).

ii) *Stereoelectronic effects.* It is essential to achieve a suitable orbital alignment for the intramolecular reaction to occur. Some quite dramatic differences in reactivity can result from small structural changes. For example treatment of the ester (**19**) with sodium methoxide gives a butyrolactone whereas the analogous reaction on ester (**20**) leads to the tetrahydropyran derivative (Figure 3.22). For a further discussion of such stereoelectronic effects in cyclisation reactions (which are highlighted in Baldwin's rules) see the references cited at the end of this chapter.

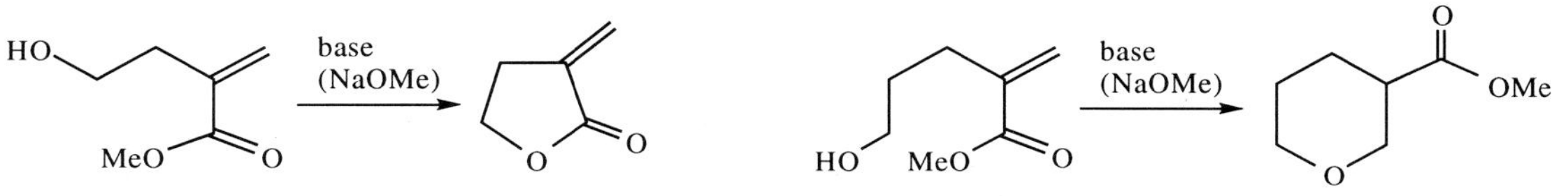

The acyloin condensation

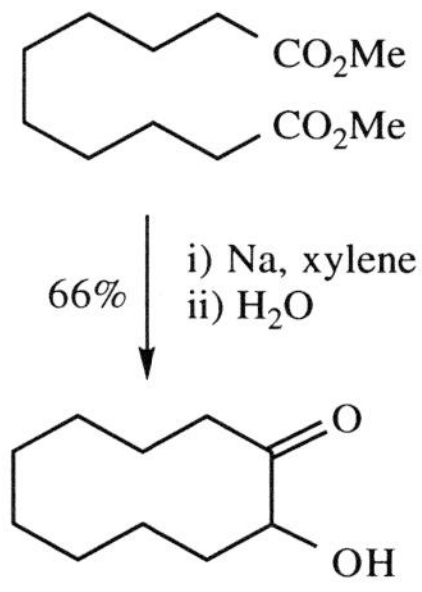

66%   i) Na, xylene   ii) H_2O

Figure 3.21

(**19**)   (**20**)

Figure 3.22 Intramolecular cyclisation reactions

iii) *Ring strain.* As a guide the strain energy in unsubstituted saturated carbocyclic rings may be ordered as follows:

small rings (3, 4) > medium rings (8–12) > common rings (5, 6, 7) = large rings (13–membered and larger).

These differences in ring strain may be used to good effect in the ring forming reaction sequence known as the Robinson annelation, e.g. in the synthesis of polycyclic ring systems such as steroids as illustrated in Figure 3.23. Reaction of the ketone (**21**) with base, e.g. sodium methoxide, generates the enolate which undergoes a Michael addition to pent–1–en–3–one to give the adduct (**22**). Reaction of (**22**) with further base also gives an enolate and examination of the structure indicates that several enolates and aldol products could potentially be formed. However only one will lead to an intramolecular aldol reaction with formation of a fused six–membered ring and indeed the product (**23**) is formed in 65% yield from the ketone (**22**).

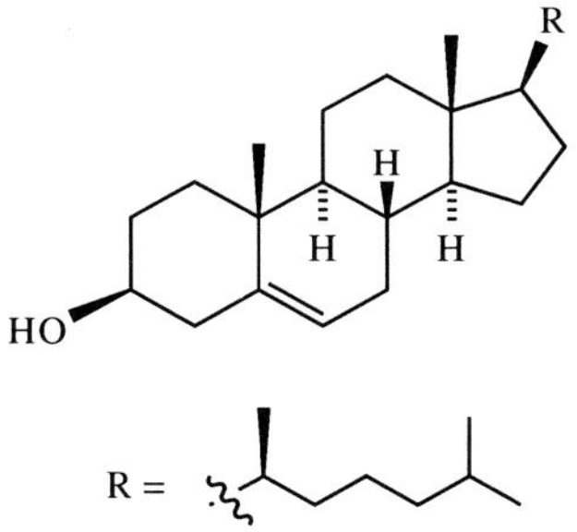

A further iteration of the annelation process may be employed for conversion of (**23**) to the tetracyclic compound (**24**). Ketone (**24**) is structurally related to the steroids, such as cholesterol, which are an important class of physiologically active compounds, and this is one of the most efficient methods for their preparation.

R =

Cholesterol

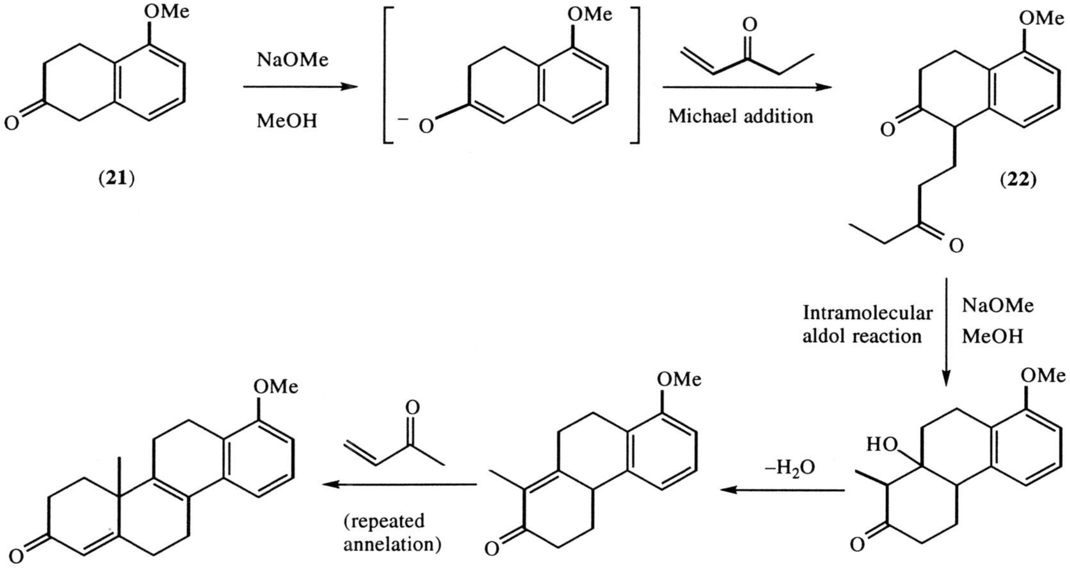

Figure 3.23 The Robinson annelation in steroid synthesis

A further important ring forming reaction which deserves particular note is the Diels–Alder reaction in which a diene and a dienophile react in a concerted process to give a six–membered ring containing a double bond. Either component can in principle contain heteroatoms and so a wide range of products containing six–membered rings may be prepared. The most successful Diels–Alder reactions are those in which an electron–withdrawing group is present on the dienophile. An electron-donating group on the diene can also be beneficial. Some examples are given in Figure 3.24, and a specific example of the use of a Diels–Alder reaction in a total synthesis is given in the next chapter.

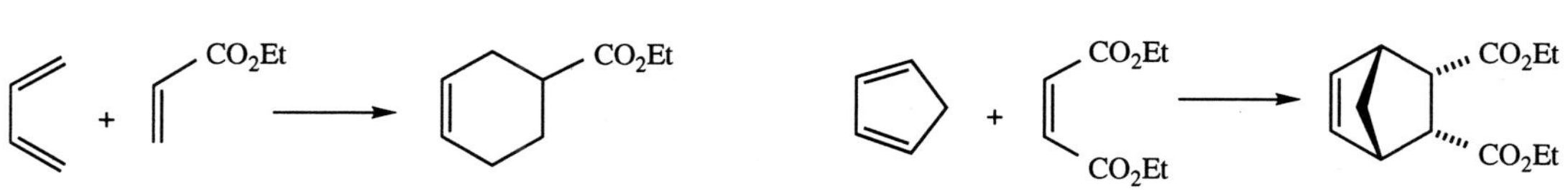

Figure 3.24 Diels–Alder reactions

3.6 Functional group interconversions (FGIs)

Most complex molecules contain many more functional groups than simply the carbonyl group and often these may be prepared from carbonyl containing functional groups. For example, consider the synthesis of the ketone (**25**), containing a double bond. Since alkenes may be prepared by the dehydration of alcohols, the first step in the retrosynthetic analysis of (**25**) could be a functional group interconversion to an alcohol. But which alcohol? In one case (**26**) the resulting substituents are in a consonant disposition whereas in the second case (**27**) they are dissonant (Figure 3.25). Obviously we can circumvent this problem by using one of the methods involving reversal of polarity described in section 3.3. However it is much simpler to use the functional group interconversion which leads to the consonant relationship of the functional groups. Further retrosynthetic analysis leads to the identification of suitable synthons and reagents. Hence the synthesis of (**25**) may be simply achieved via an aldol reaction with benzophenone and benzaldehyde.

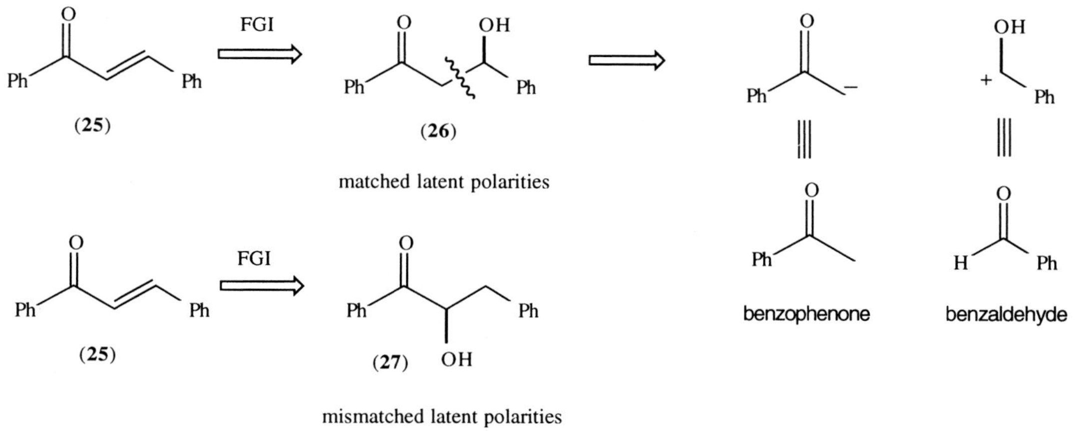

Figure 3.25 Retrosynthetic analysis of the enone (**25**)

In this section we will survey the principal functional group interconversions which are commonly encountered in organic synthesis. It is not intended to provide a list of reactions but to simply highlight the relationships between functional groups. More details of the reactions can be found in the recommended texts under further reading.

3.6.1 Functional groups containing heteroatoms

For convenience these functional groups will be initially divided into three major classes depending upon their oxidation level.

a) Carboxylic acids and their derivatives

Compounds in this class are at the highest oxidation level of organic compounds and include carboxylic acid (RCO_2H), ester/lactone (RCO_2R'), amide/lactam ($RCONHR$), anhydride ($RCO.O.COR'$), and acid chloride ($RCOCl$). They may be interconverted by a series of simple reactions as shown in Figure 3.26.

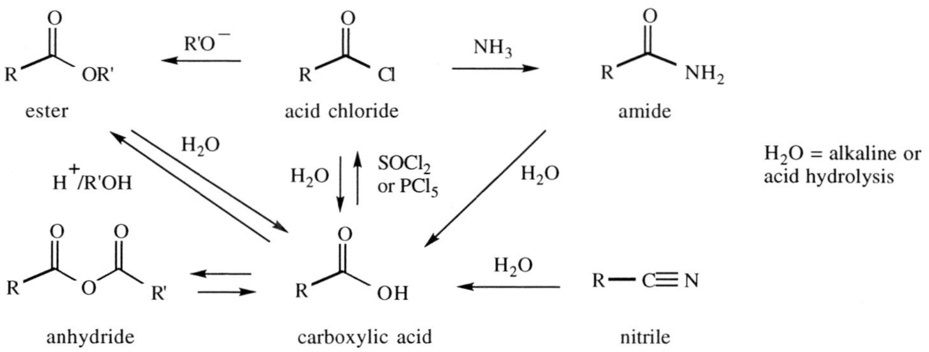

Figure 3.26 Transformations of carboxylic acid derivatives

b) Aldehydes, ketones and their derivatives

Functional groups in this class are at a lower oxidation level than class a) and may include the feature C=X in which the carbon atom is bonded directly to either hydrogen or carbon (but not to another heteroatom). The group includes aldehydes ($RHC=O$), ketones ($RR'C=O$), imines ($RR'C=NR''$), hydrazones ($RR'C=NNHR''$) and oximes ($RR'C=NOH$). These compounds may generally be interconverted using addition/dehydration reactions. For example addition of an amine to an aldehyde followed by loss of one molecule of water gives an imine (Figure 3.27).

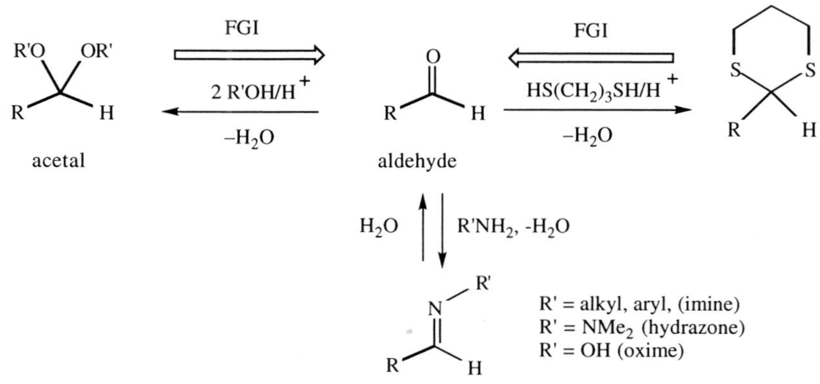

Figure 3.27 Transformations of aldehydes

The other important classes of compounds in this section are those which contain two heteroatoms attached to the same carbon atom, e.g. acetals (RO.CHR'.OR) and dithianes (RS.CHR'.SR). Methods for their interconversion are shown in Figure 3.27.

c) Alcohols and their derivatives

Apart from alcohols (ROH) themselves, this class includes amines (RNH_2), thiols (RSH), disulphides (RSSR), ethers (ROR) and alkyl halides (RX). In order to convert an alcohol to other functional groups within the same oxidation level, it is necessary to convert the hydroxyl group into a good leaving group, e.g. an alkylsulphonate ester (Figure 3.28). Sulphonate esters are prepared by the reaction of an alcohol and an appropriate alkylsulphonyl chloride in the presence of a base (n.b. the sulphonate ester is an excellent leaving group since the resultant negative charge is resonance stabilised over three oxygen atoms).

Alcohols may be converted to halides by the *in situ* activation of the hydroxyl group (Figure 3.29). A word of caution; elimination reactions can often compete with substitution to give alkenes.

Use of a sulphonate ester as a leaving group

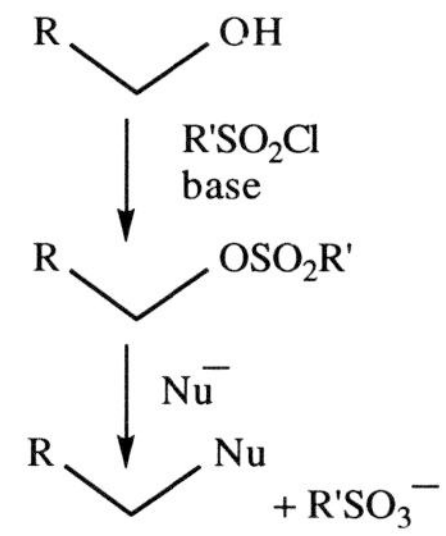

Figure 3.28

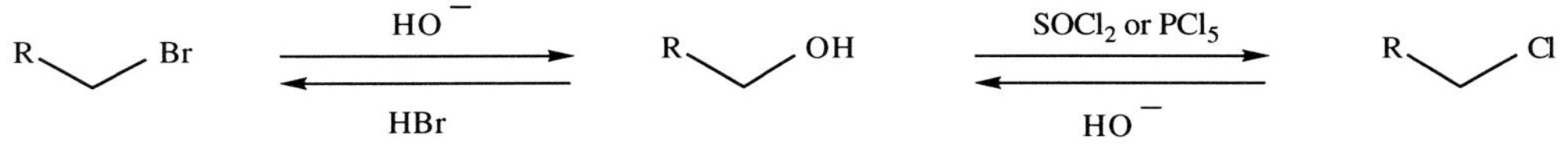

Figure 3.29 Interconversions of alcohols and halides

Interconversions between the three oxidation levels a, b, and c above

To move between groups classified in the previous section, it is necessary to perform a reduction or oxidation reaction at some stage.

Oxidation

Many methods have been developed for the oxidation of organic compounds and it is possible to transform a low oxidation level functional group into essentially any group of higher level (Figure 3.30). Further details of these reactions will be given in Chapter 5.

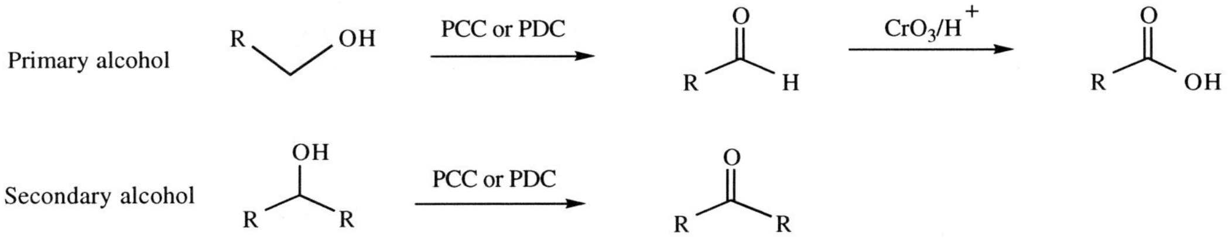

Primary alcohol

Secondary alcohol

Figure 3.30 Oxidations

Reduction

Essentially the reverse of oxidation, the reduction of carboxylic acids and their derivatives generally proceeds via the 'aldehyde oxidation level' before reaching the 'alcohol level'. Reduction of carbonyl compounds is usually achieved with metal hydride reagents, e.g. lithium aluminium hydride ($LiAlH_4$) (Figure 3.31).

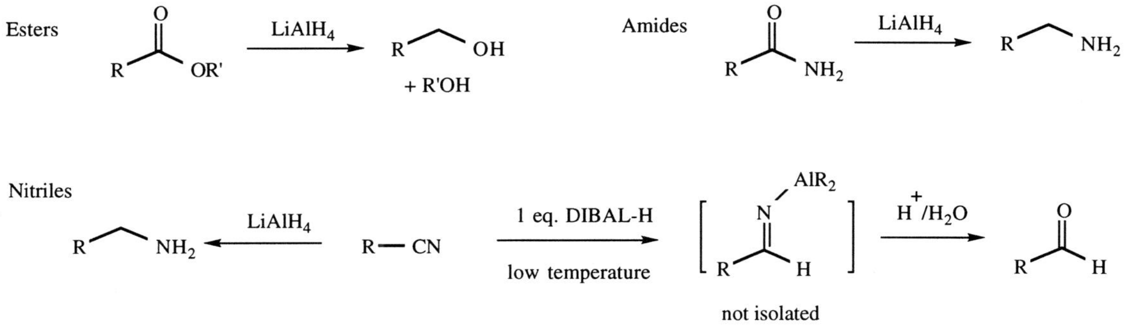

Figure 3.31 Reductions

Total removal of a functional group to form a hydrocarbon may be achieved in a number of ways. Dehydration of an alcohol followed by catalytic hydrogenation of the resultant alkene over a metal catalyst, e.g. Pd, gives the corresponding alkane. Alternatively an alcohol may be converted *either* to an alkylsulphonate ester which in turn may be reduced with $LiAlH_4$ *or* to a suitable thio–derivative (e.g. the xanthate) which may be reduced with tri–*n*–butylstannane under radical conditions to give the alkane (Figure 3.32).

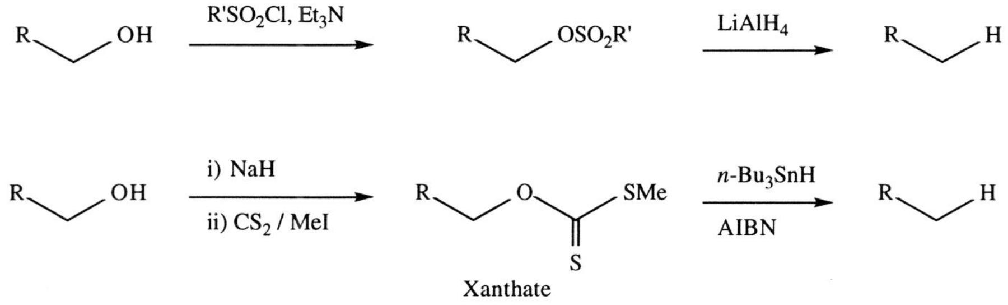

Figure 3.32 Removal of a hydroxyl group

Several methods exist for the removal of a carbonyl group including the Clemmensen reduction (Zn, HCl) and the Wolff–Kishner reduction (NH_2NH_2, KOH). The choice of reaction conditions often depends on the presence of other functionality within the molecule (Figure 3.33).

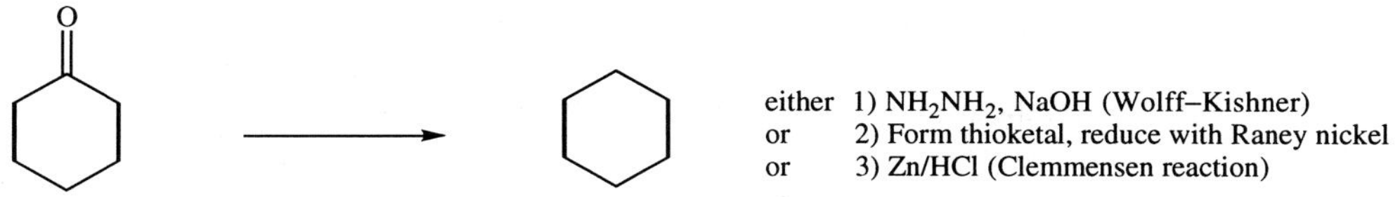

Figure 3.33 Removal of a carbonyl group

3.6.2 Unsaturated hydrocarbons

Alkenes

Alkenes are at the same oxidation level as alcohols. They may be prepared in
several ways and undergo a diverse range of functional group interconversions
as outlined in Figure 3.34. Many of the reactions represented in Figure 3.34
may be achieved with good regio– and stereo–control as discussed in later
chapters.

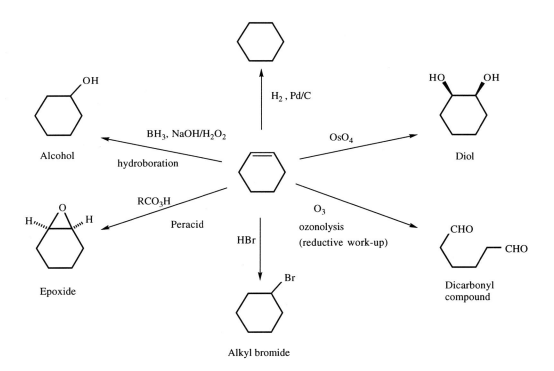

Figure 3.34 Reactions of alkenes

Aromatic compounds

Most of the chemistry of aromatic compounds involves substitution reactions in which an electrophile or nucleophile displaces a hydrogen atom or a functional group already attached to the ring. The aromaticity of the ring is maintained. However a very important reaction of aromatic rings in which the aromaticity is not retained is their reduction to 1,4–dienes using sodium in liquid ammonia (the Birch reduction, Figure 3.35). The position of the bonds relative to a substituent in the diene depends on whether the substituents are electron–withdrawing or electron–donating groups.

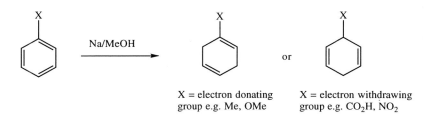

X = electron donating
group e.g. Me, OMe

X = electron withdrawing
group e.g. CO_2H, NO_2

Figure 3.35 Birch reduction of aromatic rings

3.7 Conclusion and practice examples

Most of the reactions described in this section will have been discussed in more detail in first and second year lectures and many excellent texts are available giving reaction mechanisms and experimental procedures. Armed with the knowledge of carbon–carbon bond forming reactions, functional group interconversions and the concept of polarity, the retrosynthetic approach described in this book should allow you to tackle the synthesis of some reasonably complex molecules.

It is useful at this stage to devise syntheses of a range of target molecules for yourself and we recommend that you carry out the retrosynthetic analyses and hence suggest syntheses of the following compounds from readily available starting materials. You will find the table summarising synthons and equivalent reagents given on page 13 useful when working through these examples.

from an acyclic
starting material

from an acyclic
starting material

from a cyclic
starting material

from a monocyclic
starting material

Further reading

E. J. Corey and X.-M. Cheng. *The Logic of Chemical Synthesis*, Wiley Interscience (1989).

T.-L. Ho. *Polarity Control for Synthesis*, Wiley (1991).

F. Serratosa. *Organic Chemistry in Action: The Design of Organic Syntheses*, Amsterdam (1990).

T. A. Hase. *Umpoled Synthons: A Survey of Sources and Uses in Synthesis*, John Wiley and Sons (1987).

D. Seebach. Methods of Reactivity Umpolung, *Angew. Chem., Int. Edn. Engl.,* 1979, **18**, 239.

Baldwin's rules: J. E. Baldwin, *J. Chem. Soc., Chem. Commun.,* 1976, 734.

R. C. Larock. *Comprehensive Organic Transformations*, VCH (1989).

4 Retrosynthetic analysis III: Strategy and planning

4.1 Introduction

So far in this book we have examined the basic principles of retrosynthetic analysis. Armed with this information, it should now be possible for you to devise synthetic routes to reasonably complex molecules. However there are usually a very large number of possible synthetic routes for the preparation of a given target molecule. It is clearly impractical actually to try each method in the laboratory to see which is the best as this would be costly, time consuming and frustrating. The aim of this chapter is to provide some guidelines to enable you to devise a pathway which would lead to the most efficient synthesis of the required product.

4.2 Strategy and planning

1) *Consider a range of possibilities.* Make sure that you examine a number of possible retrosynthetic approaches to the target molecule. Choose an approach for which reagents are readily available and inexpensive (refer to catalogues) and reactions which, as far as possible, give a high yield of a single product at each stage (see Chapters 5, 6 and 7 for a discussion of selectivity in synthesis).

2) *Convergent versus linear synthesis.* When considering the first disconnection in the retrosynthetic analysis try to divide the molecule into approximately equal halves, and thereafter aim to do the same with each subsequent disconnection. This will lead to a synthesis which is convergent, i.e. several 'mini–syntheses' leading to the target molecule. The alternative linear approach, in which small units are added to a growing chain, is less efficient. A schematic representation of a convergent versus linear synthesis is shown in Figure 4.1. The generalised molecule (**1**) consists of six different groups linked together in a chain. If we assume a 70% yield for each synthetic manipulation on the pathway then the linear pathway would give a 17% overall yield for the target molecule whereas the convergent approach would lead to a 34% yield (based on the longest pathway).

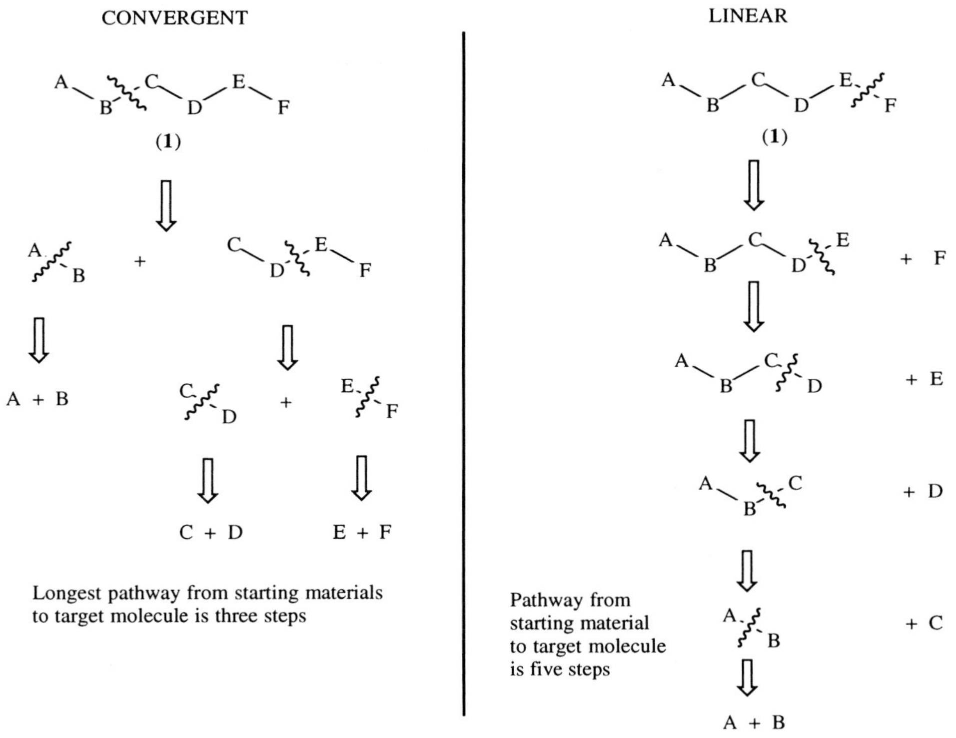
CONVERGENT
LINEAR
(1)
(1)
A + B
C + D
E + F
+ F
+ E
+ D
+ C
A + B
Longest pathway from starting materials to target molecule is three steps
Pathway from starting material to target molecule is five steps

Figure 4.1 Schematic representation of convergent and linear synthetic pathways

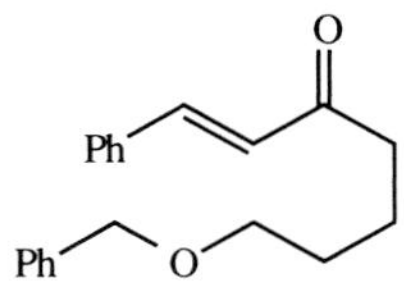
O
Ph
Ph
O

(2)

*Devise a convergent synthesis of molecule (**2**) using retrosynthetic analysis directed by the principle above.*

One notable case when a linear synthesis is effectively used is in the construction of peptides by sequentially connecting amino acids one at a time to a growing chain. This method is particularly valuable for peptide synthesis because it can be efficiently automated (see Chapter 5).

3) *Aim for greatest simplification.* Given a choice of possible disconnections, those located at branch points usually give straight chain fragments which are more likely to be commercially available or simply prepared (Figure 4.2).

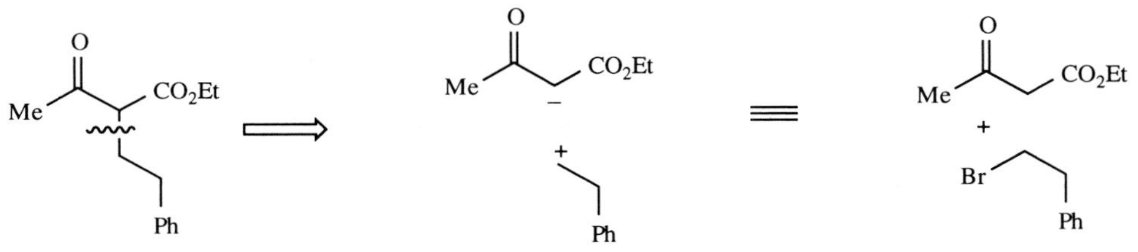
O
Me
CO_2Et
Ph
O
Me
CO_2Et
Ph
O
Me
CO_2Et
Br
Ph

Figure 4.2 Disconnection at a branch point

4) *Use any inherent symmetry in the target molecule.* The use of elements of symmetry can often dramatically reduce the number of steps involved in a synthetic pathway. In some cases the symmetry is obvious; however, often it is concealed by other structural features. For example in the target molecule (**3**) the value of symmetry may not be immediately obvious. However disconnection of the carbon–carbon bond shown in Figure 4.3 leads to the use of two molecules of cyclohexanone as starting materials for the synthesis.

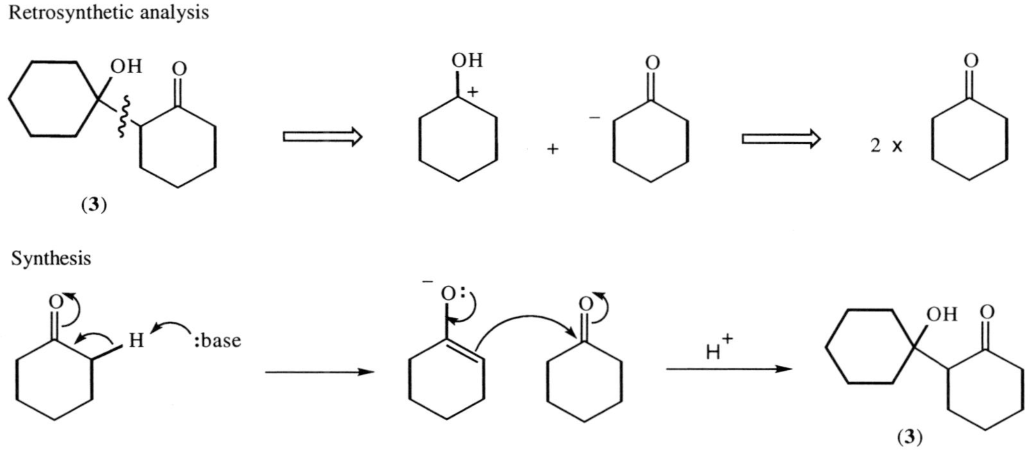

Retrosynthetic analysis

(**3**)

Synthesis

(**3**)

Figure 4.3 Retrosynthetic analysis and synthesis of (**3**)

A further example of the importance of symmetry is in the synthesis of 3–methylcyclopent–2–enone (**4**) (Figure 4.4). A functional group interconversion (FGI) of the double bond in (**4**) to the tertiary alcohol (**5**) (with a consonant pattern of latent polarities), followed by the disconnection of the carbon–carbon bond leads to the symmetrical diketone (**6**), the synthesis of which has already been described in Chapter 3.

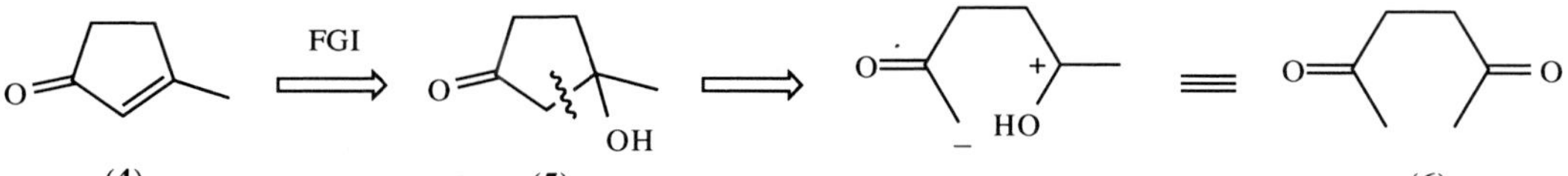

(**4**)   (**5**)   (**6**)

Figure 4.4 Retrosynthetic analysis of 3-methylcyclopent-2-enone (**4**)

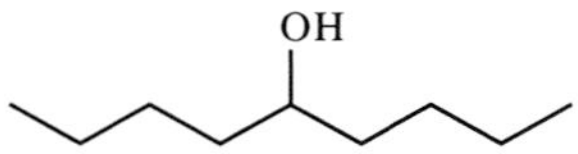

(**7**)

*Bearing in mind the importance of symmetry, use retrosynthetic analysis to devise an efficient synthesis of (**7**) using starting materials containing four carbon atoms or less.*

5) *Introduce reactive functional groups at a late stage in a synthesis.* It is often difficult to selectively react a less reactive functional group when there is more reactive functionality present within a molecule (see Chapter 5). For example an epoxide will react with a wide range of nucleophiles under both acidic and basic conditions (see Chapter 6). An elegant example of the introducion of an epoxide as the last stage of a synthesis is in the preparation of scopine (**8**) by Noyori. The ketone (**9**) was prepared in good yield using organometallic chemistry. Reduction of the ketone with DIBAL–H gave the alcohol (**10**) which on treatment with mCPBA gave the required epoxide (**8**) (Figure 4.5). The reasons for the order of the steps in this reaction sequence will become apparent in later chapters.

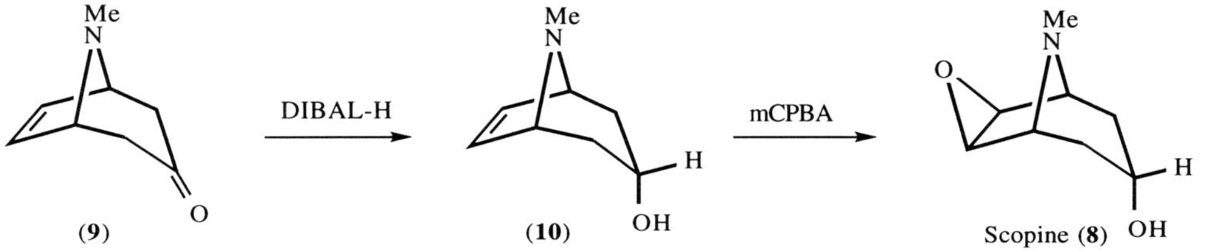

Figure 4.5 Late introduction of the reactive epoxide group in the synthesis of scopine

6) *Introduce functional groups when and where required to facilitate bond–making.* The design of a synthetic route to a target molecule must lead to the construction of the required carbon skeleton with all substituents in their correct positions and with the appropriate stereochemistry. However functional groups which are not present in the final molecule may be essential in carbon–carbon bond forming reactions. For example diethyl malonate and ethyl acetoacetate have been widely used in synthesis. The preparation of the acid (**11**) (Figure 4.6) involves the reaction of ethyl acetoacetate (**12**) with ethyl acrylate to give the diester (**13**) followed by hydrolysis and decarboxylation of the β–keto acid (**14**) (see Chapter 6).

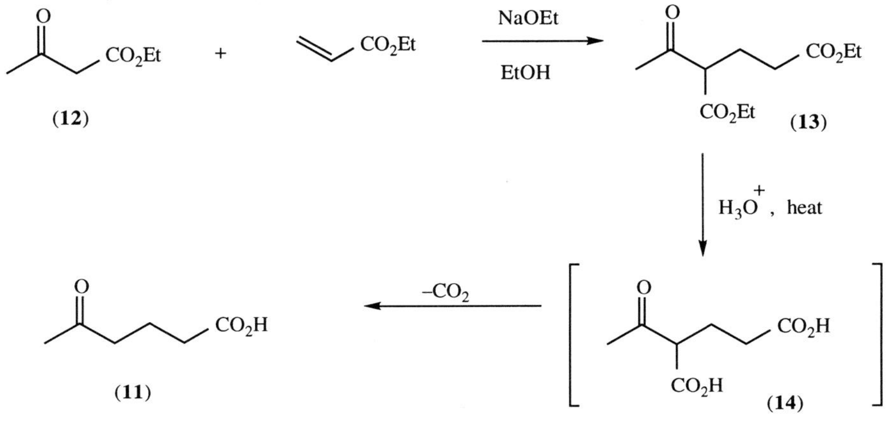

Figure 4.6 Use of an ester to facilitate carbon–carbon bond formation

If our target molecule contains a complex carbon skeleton and very few or no functional groups then clearly further functional groups will be required to assemble the requisite carbon skeleton. An elegant example of this is in the synthesis of patchouli alcohol (**15**), (Figure 4.7). An intramolecular Diels–Alder reaction of the ring diene with the terminal alkene was successfully used to establish the required carbon skeleton in one step. Catalytic hydrogenation of the resultant alkene gave the target molecule (**15**).

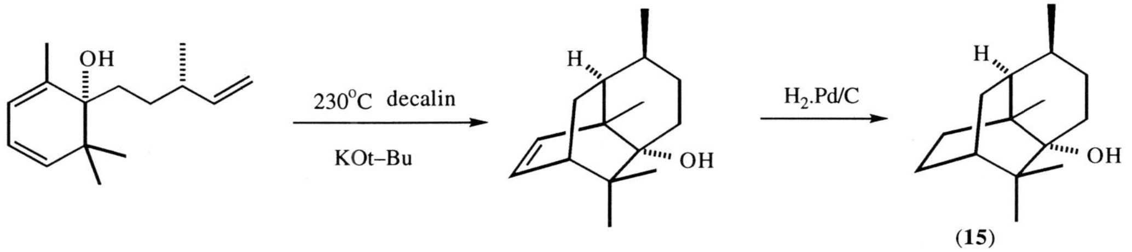

Figure 4.7 Synthesis of patchouli alcohol (**15**)

Finally

In a book of this length it is not possible to include a comprehensive coverage of all possible synthetic strategies but the basic guidelines to enable you to recognise good disconnections have been described. A more detailed discussion may be found in the books referenced at the end of Chapter 2.

Further reading

Synthesis of scopine: R. Noyori, Y. Baba and Y. Hayakawa, *J. Am. Chem. Soc.,* 1974, **96**, 3336.
Synthesis of patchouli alcohol: F. Naf and G. Ohloff, *Helv. Chim. Acta,* 1974, **57**, 1868.

5 Selectivity I: Chemoselectivity and protecting groups

5.1 Introduction

The goal in any synthesis is to obtain a pure sample of the desired product by an efficient and convenient procedure. A synthetic route to the target compound must be designed to enable the required pattern of carbon bonds to be constructed with all the functional groups and substituents in their correct positions (regiochemistry) and with control of relative orientation of the substituents (stereochemistry). In addition, the shortest route is required and each step in the synthesis should ideally give only the desired product. With the vast array of reagents and synthetic methods which are now available, it is often possible to introduce the required functionality with total regio– and stereocontrol. *Chemoselective, regioselective, stereoselective,* and *enantioselective reactions* may be used to prepare the required target compounds. These reactions may be defined as follows:

In a *chemoselective reaction* one functional group within the molecule reacts leaving further potentially reactive functionality unaffected. Reactions of this type will be discussed in this chapter.

In a *regioselective reaction* the formation of one structural (or positional) isomer is favoured (see Chapter 6).

In a *stereoselective reaction* one stereoisomer of a mixture is produced (or destroyed) more rapidly than another, resulting in the formation of a preponderance of the favoured stereoisomer (see Chapter 7).

Enantiomers are two molecules related as object to mirror image, i.e. each enantiomer of an enantiomeric pair has a non–superimposable mirror image and is described as chiral. In an *enantioselective reaction* one enantiomer is produced (or destroyed) more rapidly than the other; the amount by which one enantiomer exceeds the other in a mixture is known as the enantiomeric excess (ee), e.g. a compound containing an 85:15 ratio of enantiomers would be described as having a 70% ee, i.e. containing 30% of the racemate {15+15}, leaving 70% enantiomeric excess of one enantiomer. There is an increasing demand for the enantioselective synthesis of homochiral molecules

Regioisomers

Stereoisomers

In stereoisomers the pattern of connected bonds is identical. This is in contrast to regioisomers in which a different bond connectivity pattern is present.

Enantiomers

OH

OH

H $\cdots$ CO$_2$H   HO$_2$C $\cdots$ H

Me   Me

(*R*)-lactic acid   (*S*)-lactic acid

for evaluation as pharmaceuticals and agrochemicals. Texts are available describing the many methods now available to prepare non–racemic chiral compounds (e.g. the use of enzymes, chiral auxiliaries and asymmetric catalysts). However a discussion of enantioselective synthesis is beyond the scope of this monograph.

In the next three chapters methods will be highlighted to achieve selectivity which will enable you to use your knowledge of chemical reactions with greater effect in the design of efficient syntheses of organic molecules.

5.2 Chemoselective reactions

Many of the principal transformations involved in functional group interconversions (FGIs) were introduced in Chapter 3. The reactions may involve addition, substitution, elimination, reduction, oxidation etc.

There are a plethora of mild and selective reagents now available to effect specific transformations. For example the oxidation of a primary alcohol to an aldehyde may simply be achieved using the chromium reagents pyridinium chlorochromate (PCC) and pyridinium dichromate (PDC) or by the use of reagents based on activated sulphoxonium ion complexes, e.g. the Swern oxidation with oxalyl chloride, dimethylsulphoxide and a base such as triethylamine (Figure 5.1). Under more forcing conditions, e.g. the use of Jones reagent (CrO_3, H_2SO_4) or acidic potassium dichromate ($K_2Cr_2O_7$), it is not possible to isolate the aldehyde as it is immediately further oxidised to the carboxylic acid.

A further use of mild and selective reagents is in the reduction of carbonyl compounds. At low temperature (–78°C), DIBAL–H will reduce an ester to an aldehyde whilst under more forcing conditions, e.g. with lithium aluminium hydride ($LiAlH_4$), an ester is reduced to a primary alcohol. In contrast diborane will reduce a carboxylic acid but will leave other carbonyl containing functions, e.g. an ester, unaffected (Figure 5.2).

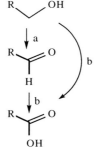

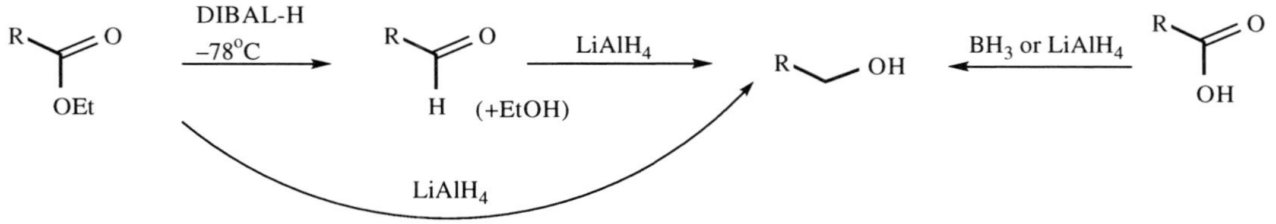

a: PCC, PDC, Swern oxidation

b: CrO_3, H_2SO_4, H_2O.

Figure 5.1 Chemoselective oxidation

Figure 5.2 Chemoselective reduction

Functional groups of different reactivity

Most molecules contain several functional groups and it is often necessary to react selectively one of these functions whilst leaving the others unchanged. As indicated above, it is often possible to achieve the required transformation

employing mild reagents in *chemoselective reactions*. Consider the reduction
of the keto–ester (**1**) shown below.

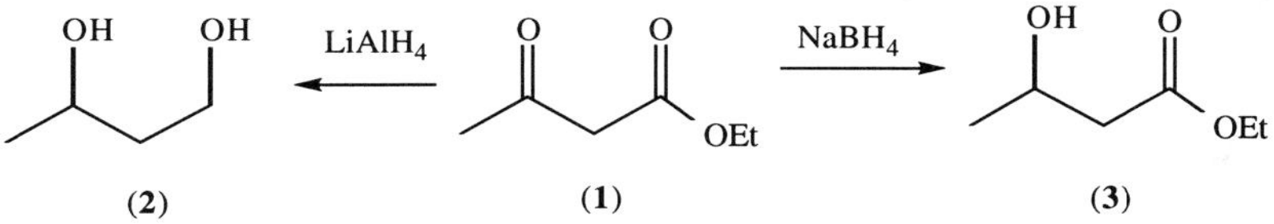

Figure 5.3 Chemoselective reduction of keto–ester (**1**)

Reaction of (**1**) with a powerful reducing agent, e.g. LiAlH$_4$, leads to
reduction of both the ketone and the ester to give a diol (**2**) whereas reduction
with a milder reagent, e.g. sodium borohydride (NaBH$_4$), gives the hydroxy
ester (**3**) through reduction of solely the ketone. A further example of a
chemoselective reduction is the reduction of an enone (Figure 5.4). Reaction
of cyclohexenone with sodium borohydride in the presence of cerium ions
gives the allylic alcohol whereas in the presence of copper(I) ions the major
product is the saturated alcohol.

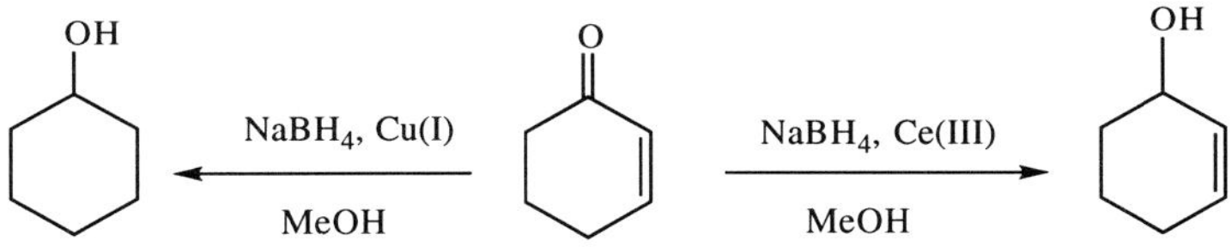

Figure 5.4 Chemoselective reduction of an enone

5.3 Protecting groups in synthesis

As a general rule, when there are two functional groups of unequal reactivity
within a molecule, the more reactive can be made to react alone. However it
may not be possible to react the less reactive functional group selectively.
For example in the case of keto–ester (**1**) shown above, the ester cannot be
directly reduced leaving the ketone unaffected. The problem may be
circumvented in two ways:

i) By the use of protecting groups.

ii) By changing the synthetic strategy to the target molecule such that it is
not necessary to react selectively a less reactive functional group.

If it is necessary to reduce an ester in the presence of a ketone, as in the
conversion of (**1**) to (**4**) (Figure 5.5), first the ketone must be converted to a
function which is stable to the reducing agent. The ester may then be reduced

and the ketone subsequently regenerated. In this case the ketone may be *protected* as the acetal (**5**).

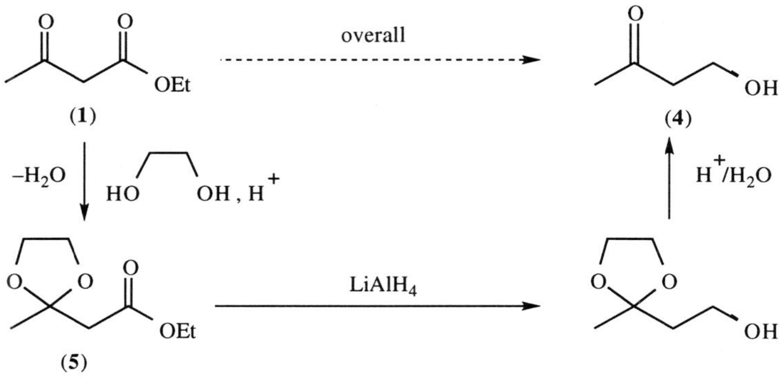

Figure 5.5   Protection of a ketone during reduction of an ester

When choosing a suitable protecting group, the following features must be considered. The protecting group must be:

i.   Simple to put on, in high yield.
ii.   Stable to the reaction conditions.
iii.   Easy to remove, also in high yield.

The acetal protecting group given in the above example satisfies all of these criteria. Protecting groups have been designed for most functional groups including alcohols, amines, carboxylic acids and alkenes. They are widely used in the synthesis of complex molecules such as peptides, amino acids, β–lactam antibiotics and carbohydrates. Excellent texts have been written describing the advantages and disadvantages of the plethora of protecting groups which are now available, hence these will not be considered in detail here. However some commonly used protecting groups for alcohols as well as the use of acetals to protect aldehydes and ketones are outlined below.

Protection of carbonyl groups as cyclic acetals
In the synthesis of cedrol (Figure 5.6) it is necessary to protect the ketone prior to reaction of the ester function with two equivalents of methyl lithium to give the tertiary alcohol. As seen in Figure 5.5 a useful protecting group for aldehydes and ketones is a cyclic acetal which may be prepared from reaction of the carbonyl compound with 1,2–ethanediol or a related diol in the presence of an acid catalyst.

The use of a diol provides an entropic advantage over the use of two equivalents of an alcohol to prepare an acetal because two molecules of starting material give two molecules of product (with a mono–ol or alcohol three molecules of starting material are required to react to form the acetal).

Cyclic acetals readily revert to aldehydes and ketones in the presence of aqueous acid. However in contrast to aldehydes and ketones, acetals are relatively inert groups. For example acetals do not react with basic or organometallic reagents (e.g. Grignard reagents) or with hydride reducing agents and hence proved valuable in the synthesis of cedrol as shown in Figure 5.6.

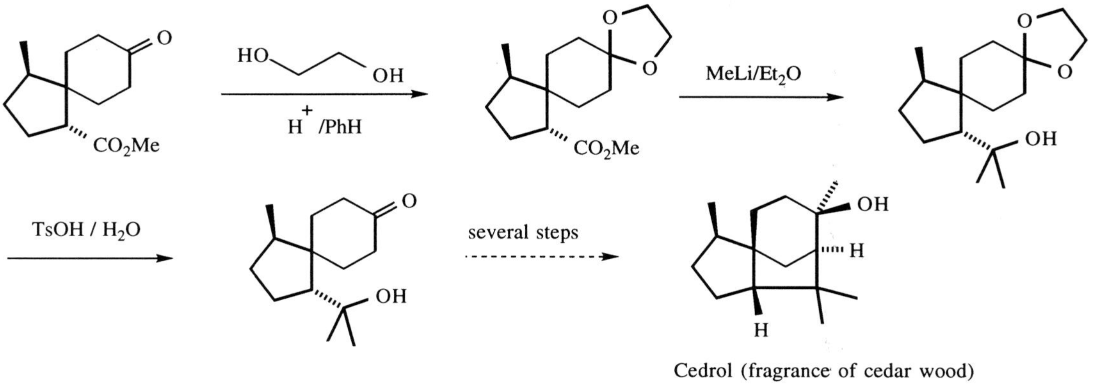

Figure 5.6 Use of an acetal as a protecting group in a natural product synthesis

Protection of alcohols

Alcohols may be converted to a range of derivatives to prevent their further reaction in synthesis. The choice of protecting group depends on the other functionality present within the molecule and the nature of the synthetic transformations to be performed.

a) Use of acetals *Acetals* not only serve as useful protecting groups for aldehydes and ketones, they also find widespread use in the protection of alcohols. Commonly used acetals include tetrahydropyranyl (THP) derivatives (prepared from dihydropyran in the presence of an acid catalyst), and the methoxyethoxymethyl (MEM) derivative (prepared by reaction of the alcohol with methoxyethoxymethyl chloride under basic conditions). Each acetal has similar properties to the cyclic acetals described above and is simply removed by treatment with aqueous acid (Figure 5.7).

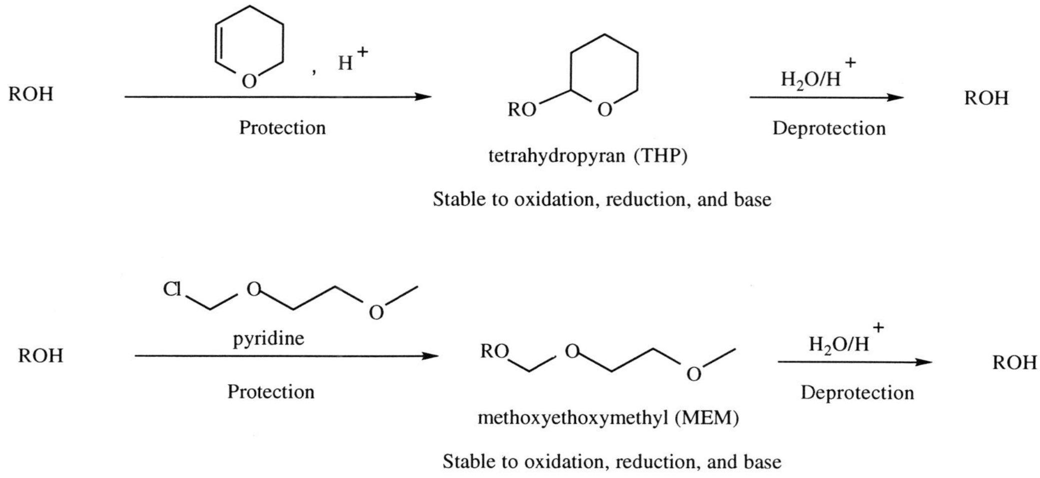

ROH $\xrightarrow[\text{Protection}]{\qquad}$ (tetrahydropyran (THP)) $\xrightarrow[\text{Deprotection}]{H_2O/H^+}$ ROH

tetrahydropyran (THP)

Stable to oxidation, reduction, and base

ROH $\xrightarrow[\text{Protection}]{\text{pyridine}}$ (methoxyethoxymethyl (MEM)) $\xrightarrow[\text{Deprotection}]{H_2O/H^+}$ ROH

methoxyethoxymethyl (MEM)

Stable to oxidation, reduction, and base

Figure 5.7 Acetals as protecting groups for alcohols

b) Use of ethers Alcohols may be simply converted to *ethers* by nucleophilic attack on a suitable alkyl halide. Ethers are stable to basic and mildly acidic conditions; they do not react with oxidising or reducing agents and are inert to organometallic reagents. However this stability means that many ethers are not easily cleaved to their parent alcohol under mild conditions. Therefore only certain ethers are commonly used as protecting groups for alcohols, e.g. benzyl ethers (Figure 5.8) which are converted to alcohols under neutral conditions by catalytic hydrogenolysis and *t*–butyl ethers (Figure 5.8) which are readily hydrolysed with dilute acid.

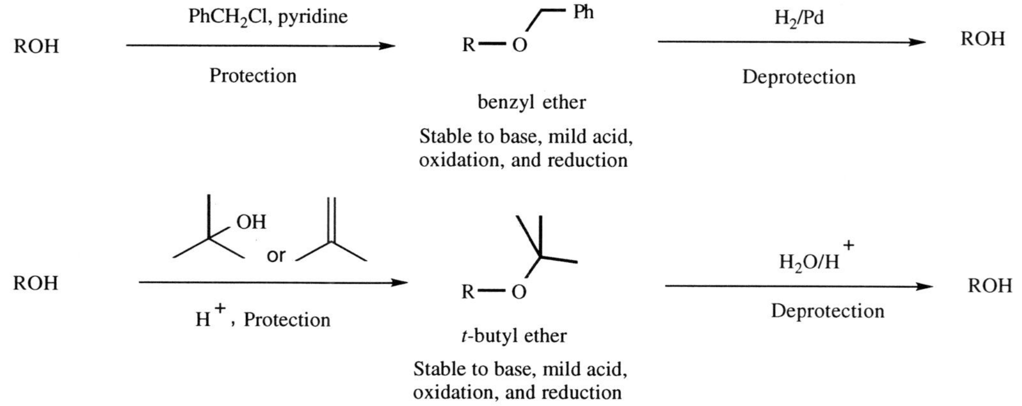

ROH $\xrightarrow[\text{Protection}]{\text{PhCH}_2\text{Cl, pyridine}}$ R—O—CH$_2$Ph $\xrightarrow[\text{Deprotection}]{H_2/Pd}$ ROH

benzyl ether

Stable to base, mild acid,
oxidation, and reduction

ROH $\xrightarrow[\text{H}^+\text{, Protection}]{\qquad}$ R—O—C(CH$_3$)$_3$ $\xrightarrow[\text{Deprotection}]{H_2O/H^+}$ ROH

t-butyl ether

Stable to base, mild acid,
oxidation, and reduction

Figure 5.8 Ethers as protecting groups for alcohols

A *t*–butyl ether protection step has been employed in the series of reactions shown in Figure 5.9, in which the positions of a ketone and an alcohol are transposed. The initial protection permits the differentiation of two alcohol groups, one of which must be subsequently oxidised.

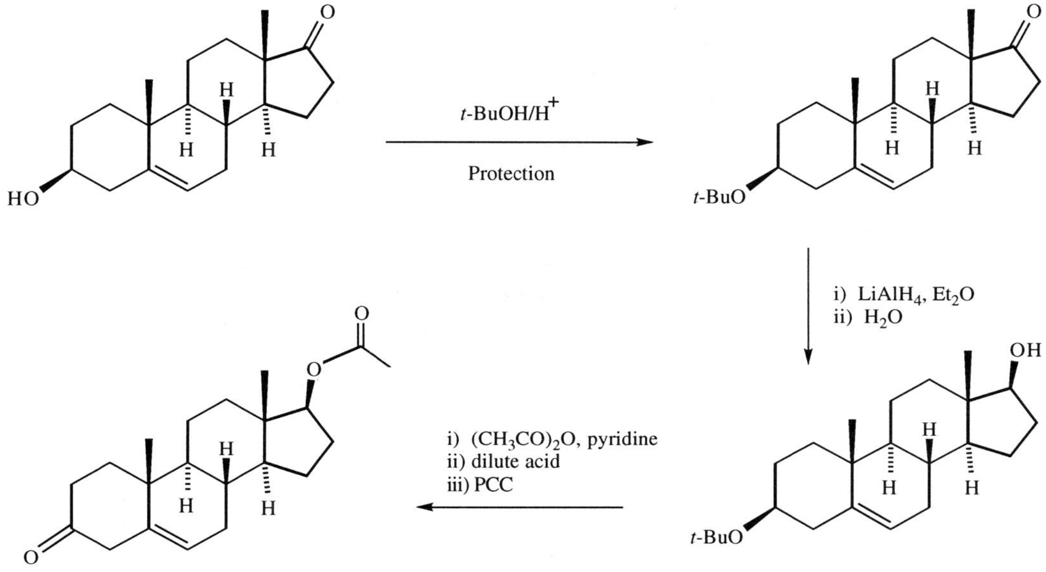

Figure 5.9 Use of protecting groups in functional group manipulation of a steroid

c) Use of trialkylsilyl ethers A further group which has found widespread use for the protection of alcohols in organic synthesis, are *trialkylsilyl ether* derivatives such as trimethylsilyl (TMS) ethers. These are prepared by reaction of an alcohol with trimethylsilyl chloride in the presence of a base. However trimethylsilyl ethers are not particularly stable and are cleaved under acidic or basic conditions and are attacked by certain nucleophiles (Figure 5.10).

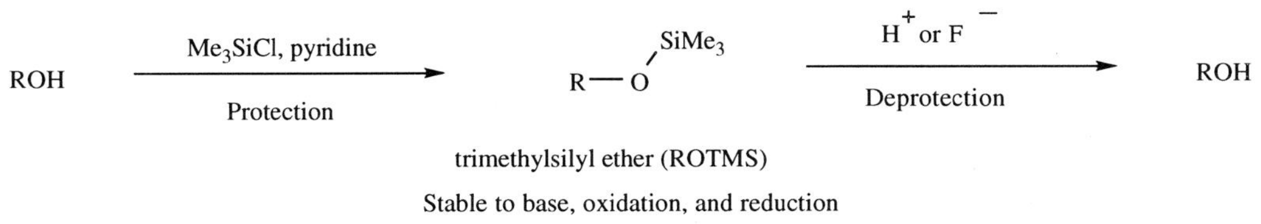

trimethylsilyl ether (ROTMS)

Stable to base, oxidation, and reduction

Figure 5.10 Use of a trimethylsilyl ether to protect alcohols

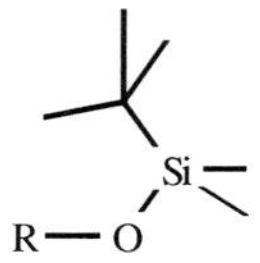

t-Butyldimethylsilyl ether
(TBDMSOR)

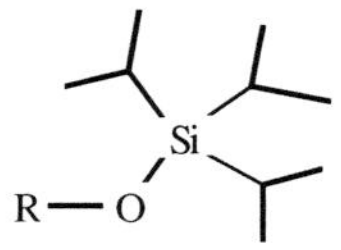

Triisopropylsilyl ether
(TIPSOR)

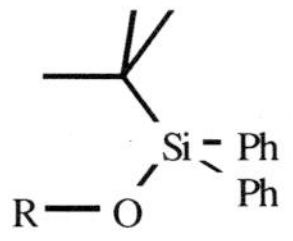

t-Butyldiphenylsilyl ether
(TBDPSOR)

To overcome these problems a number of more bulky trialkylsilyl protecting groups have been developed, e.g. *t*–butyldimethylsilyl (TBDMS), triisopropylsilyl (TIPS), and *t*–butyldiphenylsilyl (TBDPS) ethers. TBDMS ethers are more stable to hydrolysis than TMS ethers by a factor of 10^4. The steric bulk which renders the TBDMS group more stable than the TMS group also hinders the formation of the ether from *t*–butyldimethylchlorosilane. However reaction of the alcohol with the silyl chloride in the presence of imidazole gives high yields of the silyl ether.

Silyl ethers may be cleaved to their parent alcohols by fluoride ion. A convenient source of fluoride ion for this application is tetra–*n*–butylammonium fluoride (Bu4NF, or TBAF). The use of hydrofluoric acid itself is not favoured due to its corrosive nature.

In the synthesis of leukotriene B4 reported by Corey, a key early intermediate (**6**) containing both silyl and acetal protecting groups was prepared using a Wittig reaction (see Chapter 6). Using mild acid conditions it was possible to selectively hydrolyse the acetal without removing the silyl group to give (**7**), thus freeing one hydroxy group selectively for use in another Wittig reaction. The silyl group was kept in place until the end of the synthesis, which is shown in abbreviated form in Figure 5.11.

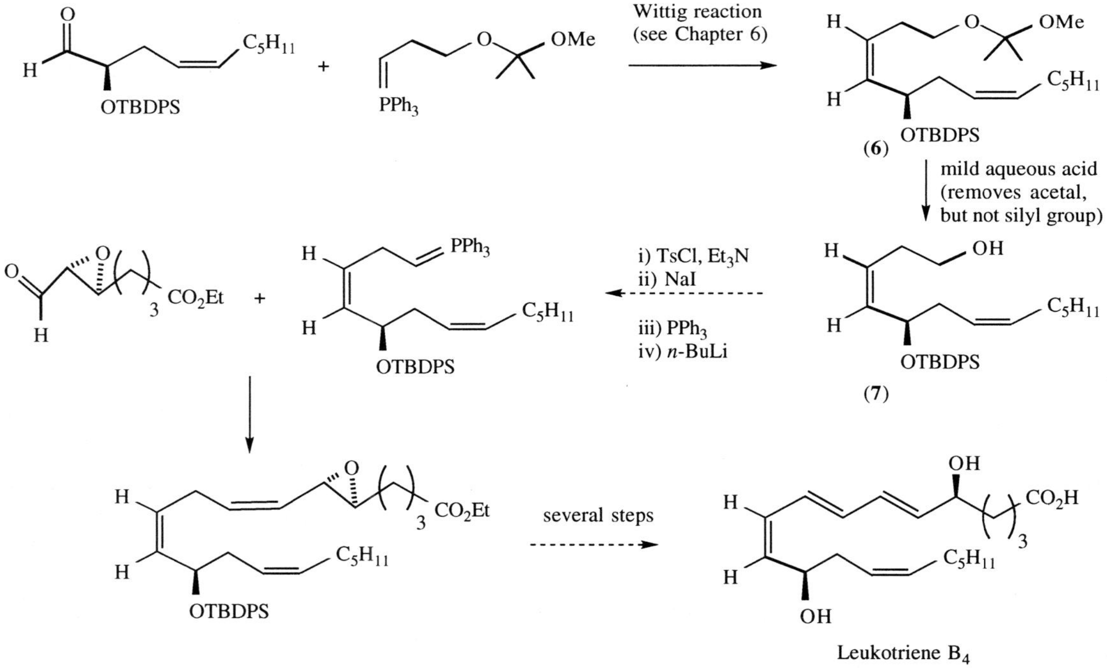

Figure 5.11 Synthesis of leukotriene B4

Protection of amines

The reactive lone pair of electrons on the amino group allows protonation of an amine as well as reactions with electrophiles. It is therefore often necessary to protect this group in a form in which the lone pair is much less reactive. The most efficient and convenient way of doing this is to convert the amine into an amide or a carbamate group. The carbonyl group effectively withdraws electron density from the nitrogen and renders it unreactive. When required, the amine can be regenerated by acid catalysed hydrolysis of the amide or carbamate. The *t*–butylcarbamate group is especially valuable in this respect since it is easily removed using mild acid. It has been successfully exploited in the Merrifield process for the synthesis of peptides in which a peptide chain is 'grown' one amino acid at a time from a solid polymer support (Figure 5.12). After all the amino acids have been added the peptide is removed from the polymer by hydrolysis.

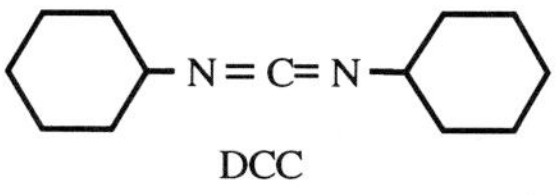

DCC is dicyclohexylcarbodiimide, a versatile dehydrating agent

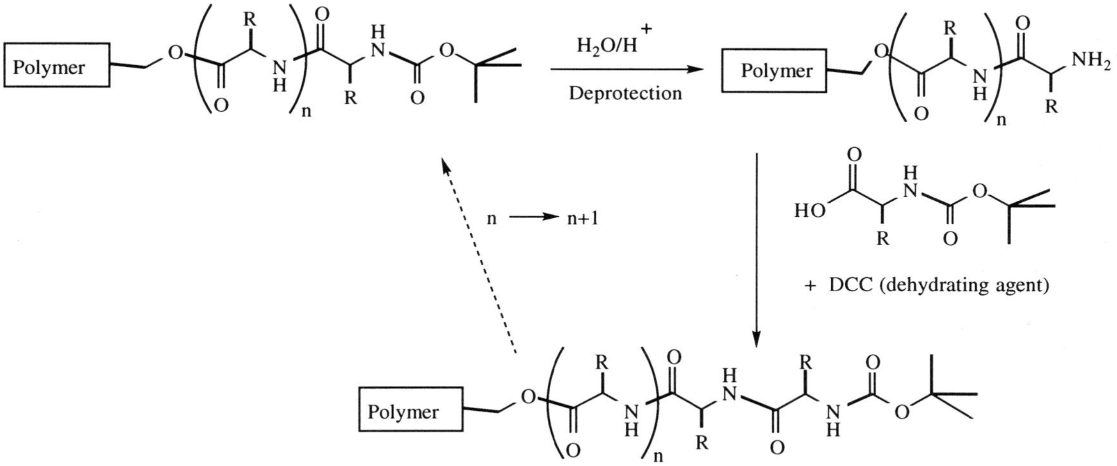

Figure 5.12 Construction of a peptide chain using the Merrifield process

Avoiding the use of a protecting group

Although protecting groups should be easy to put on and readily cleaved in high yield, one of the criteria of a good synthetic strategy is to keep the route as short as possible. Therefore ideally a synthetic scheme should be designed and reagents chosen such that the use of protecting groups is kept to a minimum.

As an example consider the conversion of 3–bromopropan–1–ol to 3–deuteriopropan–1–ol. It is well established that hydrolysis of a Grignard reagent with D_2O is an effective method for the introduction of deuterium into a molecule. If this reaction is to be utilised in this case, the alcohol must be protected, for example as the *t*–butyl ether derivative, prior to formation of the Grignard reagent. (*Why is it necessary to protect the alcohol prior to formation of the Grignard reagent?*) Reaction of the O–protected

In the radical reaction azo(iso–bis butyronitrile) (AIBN) acts as an initiator by forming radicals as shown below. These initiate a radical chain reaction between the tin hydride and the bromide, resulting in reduction.

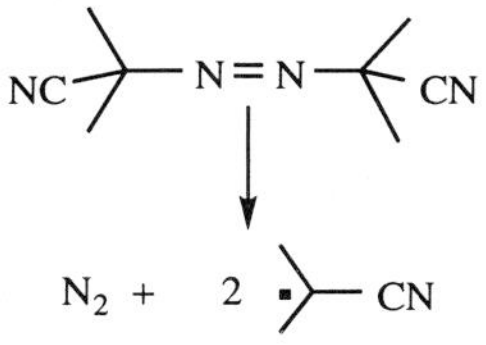

Grignard reagent with D_2O followed by deprotection of the ether with aqueous sulphuric acid would then give 3–deuteriopropan–1–ol.

However a more direct route to 3–deuteriopropan–1–ol from 3–bromopropan–1–ol which does not require protection of the alcohol, is the radical reduction of the bromide with tri–*n*–butyl–[²H]–stannane in the presence of an appropriate radical initiator (Figure 5.13).

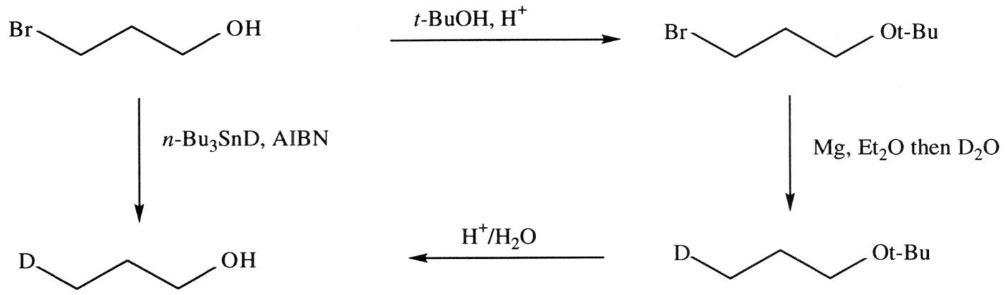

Figure 5.13 Synthesis of 3–deuteriopropan–1–ol

5.4 Reaction of just one of two identical functional groups

When a molecule contains two reactive groups, we have demonstrated that it is possible to selectively react each functional group separately with the use of mild reagents and/or protecting groups. However not only is it possible to differentiate between two functional groups of unequal reactivity within a molecule, it may also be possible to react just one of two identical functional groups. This may be achieved with limited success by reaction with one equivalent of reagent, e.g. in the conversion of the diol to the ethyl ether (Figure 5.14). The other compounds which are present in the reaction mixture are the diol and the diether; these are easily separable. However in the case of a molecule with two *almost identical* groups, reaction with one equivalent of reagent would lead to a mixture of products which would be difficult to separate.

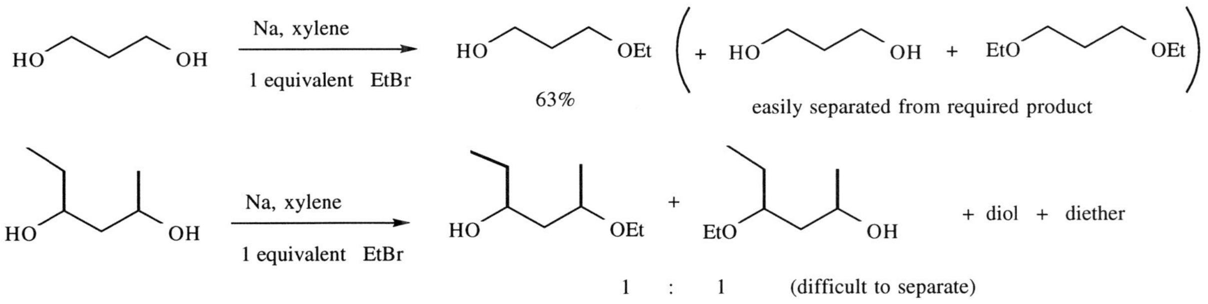

Figure 5.14 Monoalkylation of a diol using one equivalent of ethyl bromide

Alternatively only one of two identical groups may react if the product is less reactive than the starting material, as for example in the partial reduction of 1,3–dinitrobenzene with sodium hydrogen sulphide. In this example the increased electron density in the aromatic ring of 3–nitroaniline reduces the reactivity of the nitro group.

A more reliable method is to use a derivative of the identical functional groups which may react selectively as for example in the selective reduction and esterification of the 19–carboxylic acid of the fungal product fujenal diacid. The diacid is converted to the anhydride (**8**) prior to reduction with sodium borohydride to give the lactone (**9**) or formation of the ester (**10**) with sodium methoxide (Figure 5.15).

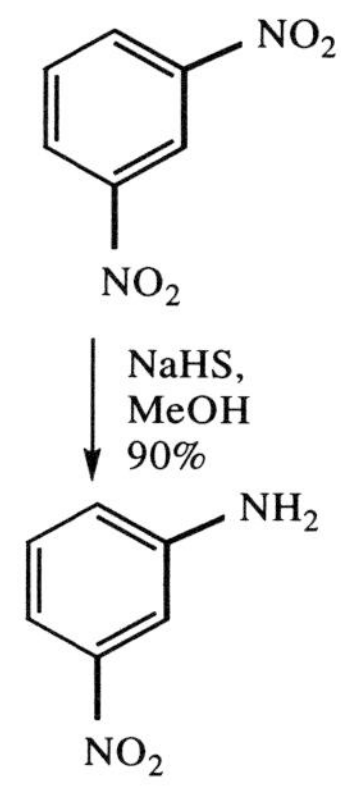

Selective reduction by formation of a less reactive product

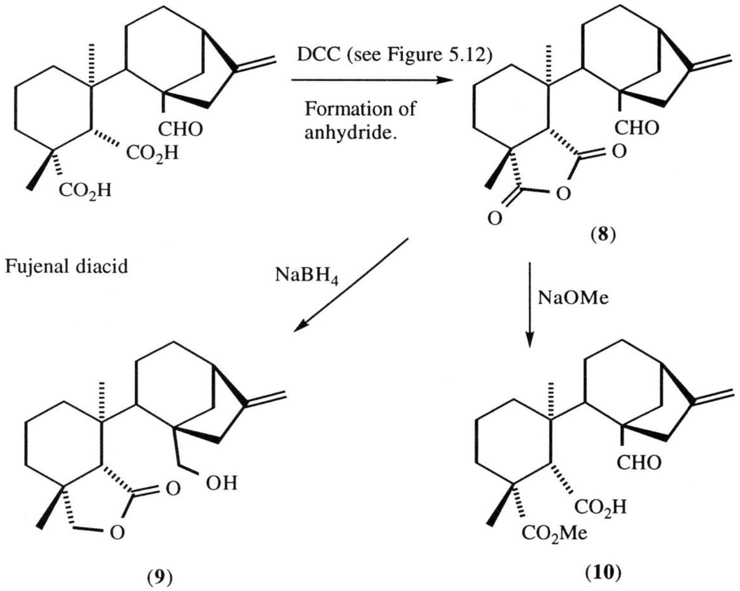

Figure 5.15 Chemoselectivity in natural product synthesis

5.5 Functional groups which can react twice

Many functional groups are capable of reacting more than once. An example of this is the amino group, which may be alkylated several times and often gives a mixture of products (Figure 5.16).

$$RX \ + \ NH_3 \longrightarrow RNH_2 \ + \ R_2NH \ + \ R_3N \ + \ R_4\overset{+}{N} \ \overset{-}{X}$$

Figure 5.16 Amine alkylation

The way round this problem is to *acylate* the nitrogen atom. The resultant amide is not capable of further alkylation. Reduction to an amine with lithium aluminium hydride then gives the required monoalkylated product (Figure 5.17).

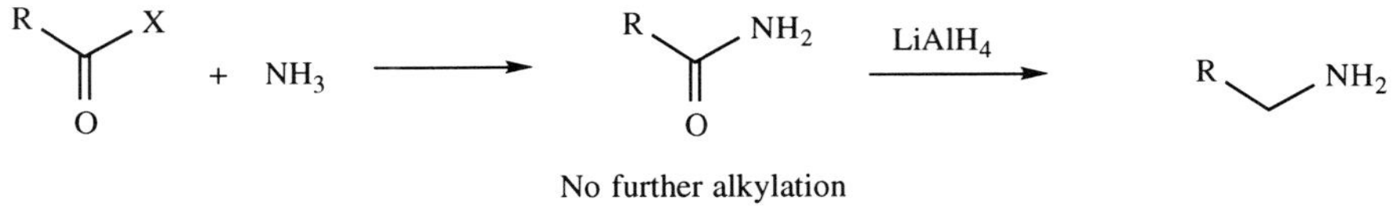

No further alkylation

Figure 5.17 Amine alkylation strategy

An alternative approach for the preparation of a primary amine is via the alkylation of phthalimide (**11**). Reaction of the alkylated product (**12**) with hydrazine results in release of the corresponding primary amine. This sequence is known as the Gabriel synthesis of an amine (Figure 5.18).

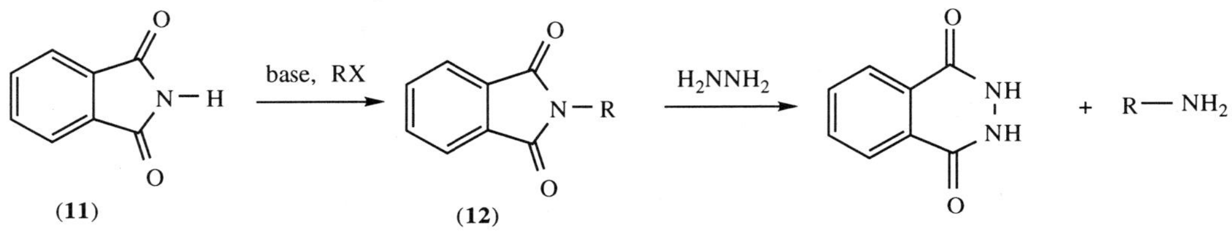

(**11**)   (**12**)

Figure 5.18 The Gabriel synthesis of an amine

Another example of chemoselective control is observed in the electrophilic alkylation of aromatic rings. In principle benzene may be substituted six times; however, monoalkylation can be achieved by forming a product which is less reactive than the starting material. Friedel–Crafts acylation and nitration are examples of valuable aromatic substitution reactions which satisfy this criterion (Figure 5.19). For more information on the mechanism and applications of aromatic reactions we recommend the OUP primer *Aromatic Chemistry*, number 4 in this series.

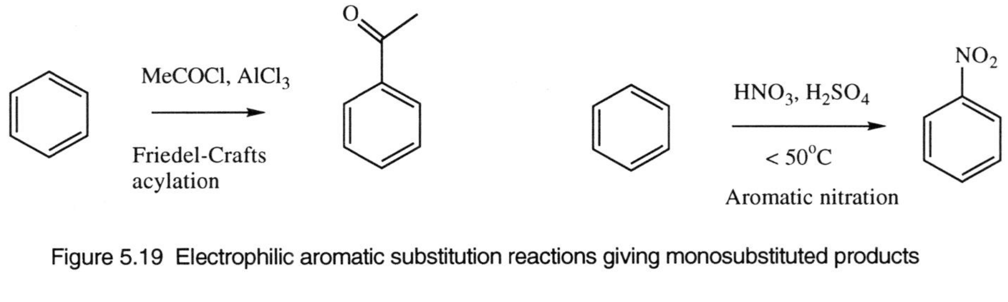

MeCOCl, AlCl$_3$

Friedel-Crafts acylation

HNO$_3$, H$_2$SO$_4$

< 50°C

Aromatic nitration

Figure 5.19 Electrophilic aromatic substitution reactions giving monosubstituted products

5.6 Practice examples

Using retrosynthetic analysis, devise syntheses of the following molecules.
To direct your approach, starting materials are suggested in each case.

Further reading

T. W. Greene and P. G. M. Nuts. *Protective Groups in Organic Synthesis*
(2nd edn), John Wiley and Sons (1991).
M. Sainsbury. *Aromatic Chemistry* (Oxford Chemistry Primer no. 4),
Oxford University Press (1992).

References to featured syntheses
Synthesis of cedrol: E. J. Corey, N. N. Girotra and C. T. Mathew, *J. Am.
Chem. Soc.,* 1969, **91**, 1557.
Synthesis of leukotriene LTB$_4$: E. J. Corey, P. B. Hopkins, J. E. Munroe,
A. Marfat and S.–I. Hashimoto, *J. Am. Chem. Soc.,* 1980, **102**, 7986.

6 Selectivity II: Regioselectivity

6.1 Introduction

When devising a synthetic route to an organic molecule, substituents and functional groups must be placed in the required positions, i.e. with the correct regiochemistry. In this chapter we begin by evaluating the regioselectivity of some of the most common methods for the preparation of alkenes. Later the control of the reactivity of functional groups which are capable of giving more than one product will be described, with particular reference to:

a) Electrophilic addition to a double bond (section 6.3)

and/or

b) Electrophilic aromatic substitution (section 6.4)

and/or   and/or

c) Electrophilic addition to an enolate (section 6.5)

i) base
ii) R^2X

and/or

d) Nucleophilic addition to an enone (section 6.6)

:Nu

and/or

e) Nucleophilic addition to an epoxide (section 6.7)

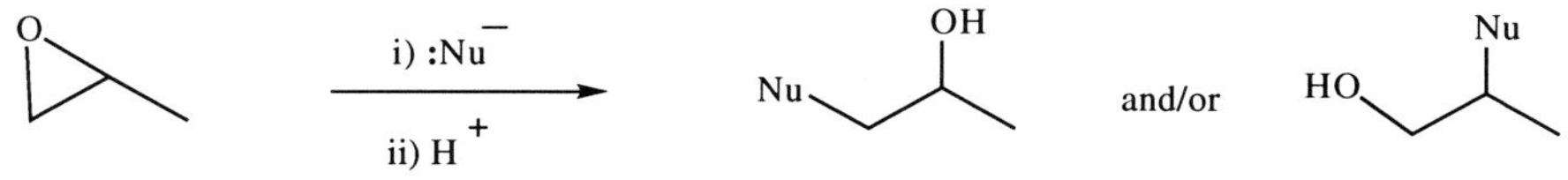

f) Oxidation of a ketone to an ester or lactone (section 6.8)

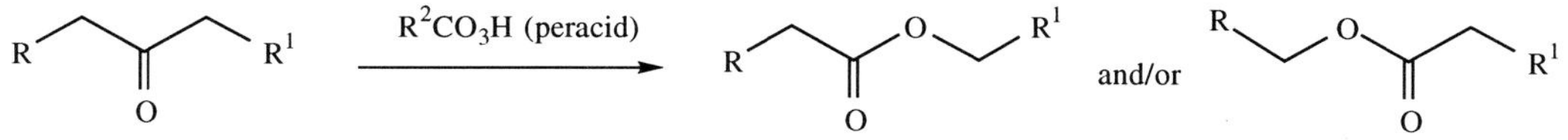

6.2 Methods for the preparation of alkenes

Consider for example the total synthesis of muscalure, the sex attractant of the female house–fly *Musca domestica.* The design of a synthetic route to this molecule must take into account the formation of a *cis* double bond between carbons 9 and 10.

Which of the many known methods to form alkenes would be most suitable in this case? The merits of three methods are considered below.

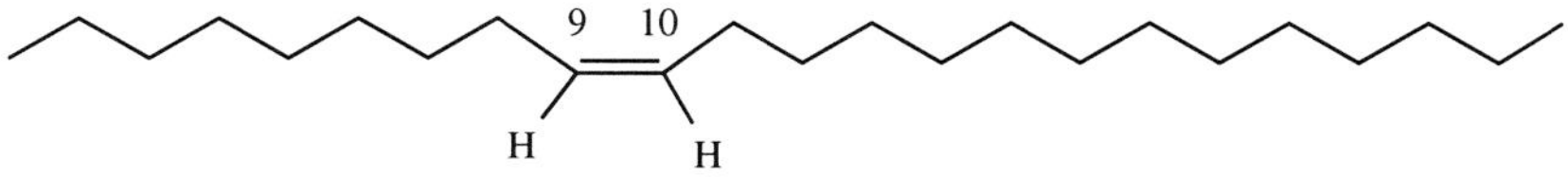

Muscalure

1. Dehydration of alcohols

Perhaps the most obvious way of making alkenes is via alcohols which undergo dehydration when heated with strong acids. One of the advantages of this approach is that alcohols are easy to prepare, e.g. by the addition of Grignard reagents (RMgBr) to carbonyl compounds or the reduction of carbonyl compounds (Figure 6.1).

However the reaction is of limited use since mixtures of alkenes are usually formed which may be difficult to separate by either distillation or chromatography. The product with the more substituted double bond generally predominates (Saitzeff's rule), e.g. the acid catalysed dehydration of 2,3-dimethylbutan-2-ol gives a 9:1 mixture of two alkenes (**1**) and (**2**) (Figure 6.2).

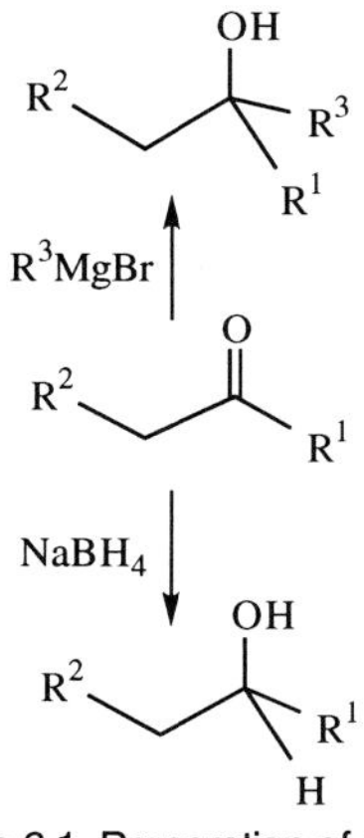

Figure 6.1 Preparation of alcohols

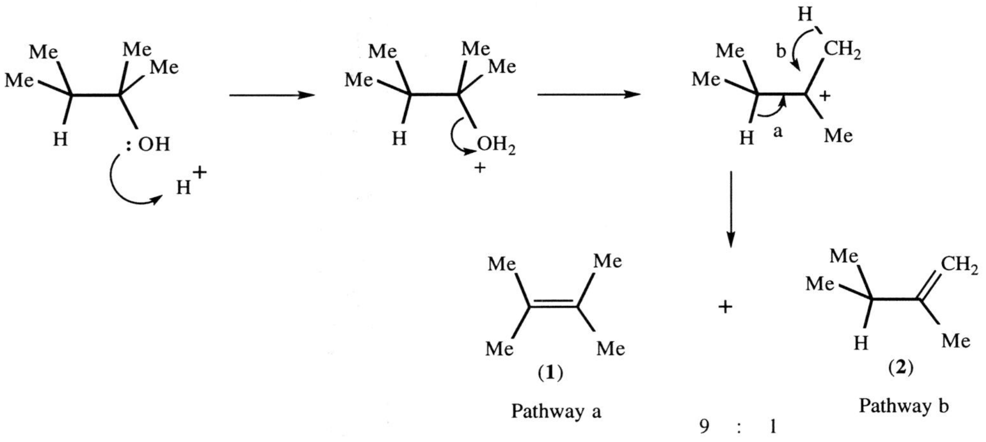

Figure 6.2 Dehydration of 2,3–dimethylbutan–2–ol

A further disadvantage of this approach is that rearrangements of the carbon skeleton may occur under the acidic conditions. For example treatment of the alcohol (**3**) with acid leads to migration of a methyl group to generate the more stable tertiary carbocation (**5**) from (**4**) before subsequent deprotonation to the final products (Figure 6.3).

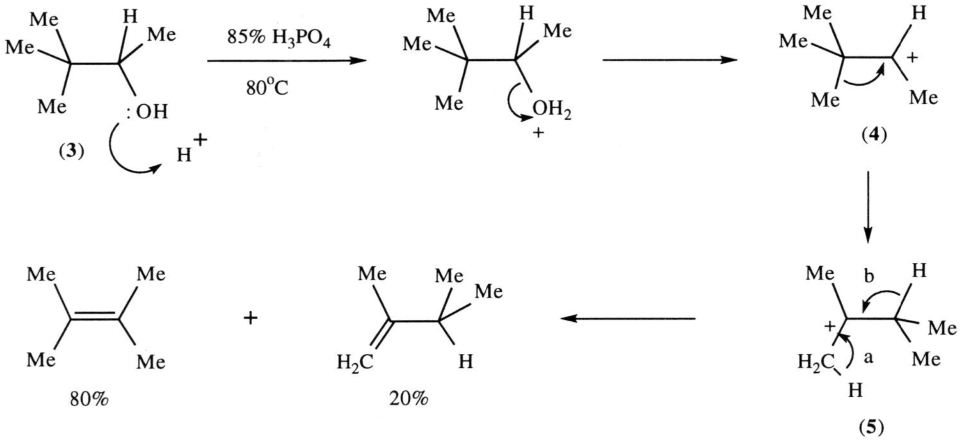

Figure 6.3 Acid catalysed migration of an alkyl group

An important use of the dehydration reaction is in the synthesis of α,β–unsaturated carbonyl compounds. Initially a new carbon–carbon bond is formed between two carbonyl containing molecules (aldol condensation) and is followed by dehydration of the resultant β–hydroxy derivative (**6**) (Figure 6.4).

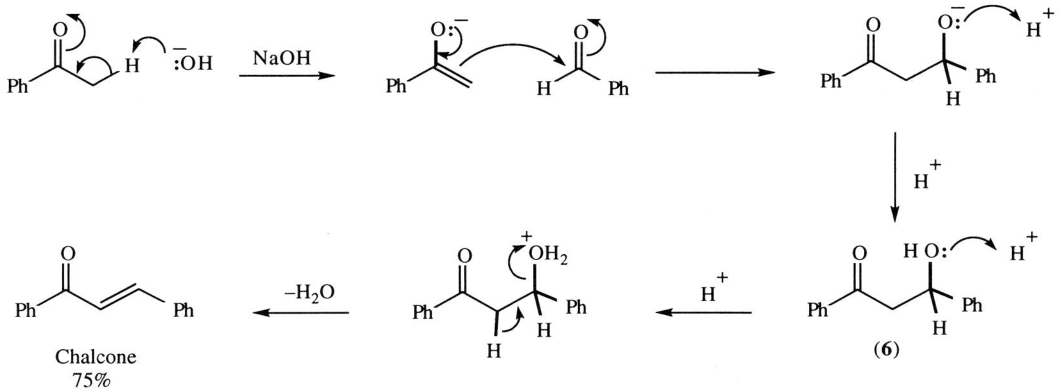

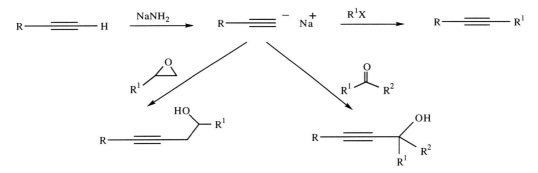

Figure 6.4 The aldol condensation

2. Reduction of alkynes

A diverse range of compounds incorporating an alkyne may be prepared using the acetylenic anion as a nucleophile (Figure 6.5).

Figure 6.5 Preparation of alkynes

Reduction of the alkyne is then an excellent method for the regiospecific preparation of alkenes since the position of the alkene is fixed from the parent alkyne. In addition the geometry of the double bond may be controlled by the choice of reducing conditions. *Cis* alkenes are formed with hydrogen in the presence of Lindlar's catalyst whereas alkynes react with sodium and liquid ammonia to give *trans* alkenes (Figure 6.6).

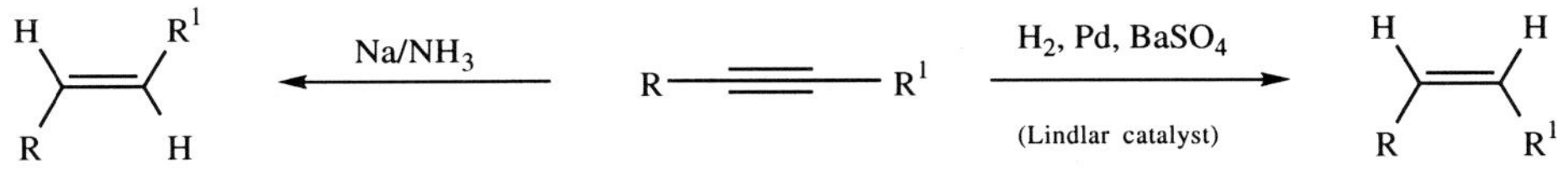

Figure 6.6 Reduction of alkynes

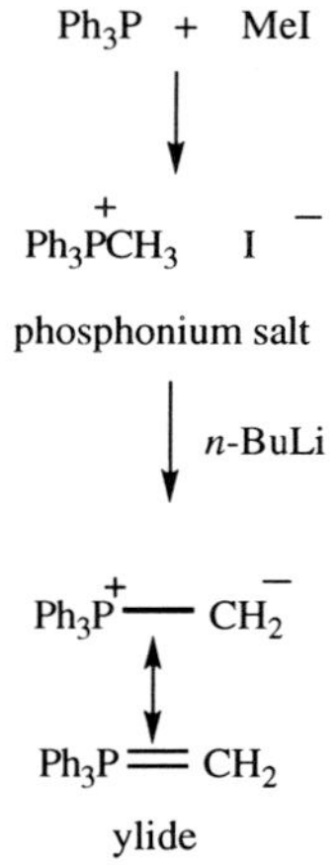

Ph$_3$P + MeI

$$\downarrow$$

Ph$_3$$\overset{+}{P}CH_3$ I$^-$

phosphonium salt

$$\downarrow \quad n\text{-BuLi}$$

Ph$_3$$\overset{+}{P}$——$\overset{-}{C}H_2$

$$\updownarrow$$

Ph$_3$P══CH$_2$

ylide

Preparation of phosphorus ylide

3. Use of phosphorus ylides–the Wittig reaction

The Wittig reaction is an exceedingly useful method of preparing alkenes. An alkyl phosphonium salt, prepared from an alkyl halide and triphenylphosphine, is deprotonated with a strong base such as *n*–butyl lithium to give an ylide. Most primary and secondary alkyl halides give good yields of phosphonium salts. The ylide then reacts with aldehydes and ketones to give an alkene with complete regiospecificity, the position of the double bond being fixed by the position of the original carbonyl function (Figure 6.7).

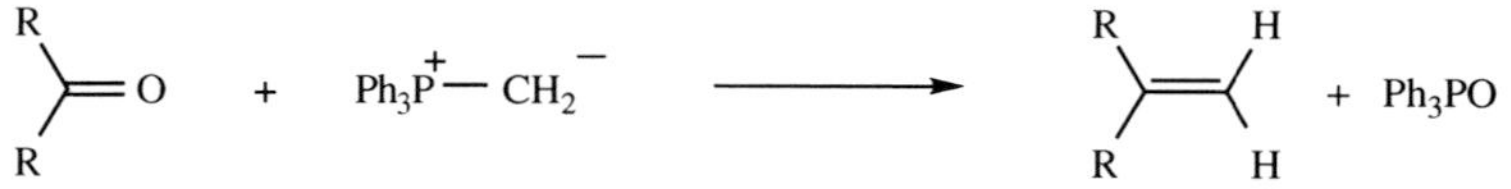

$$
\underset{R}{\overset{R}{>}}{=}O \;+\; Ph_3\overset{+}{P}{-}\overset{-}{C}H_2 \;\longrightarrow\; \underset{R}{\overset{R}{>}}{=}\underset{H}{\overset{H}{<}} \;+\; Ph_3PO
$$

Figure 6.7 The Wittig reaction.

A mixture of *cis* and *trans* isomers may be formed in the Wittig reaction but as a general rule unstabilised ylides (i.e. Ph3P=CHR, where R is an alkyl group) give mainly *cis* double bonds with aldehydes. Stabilised ylides (i.e. Ph3P=CHR, where R is an ester (CO2R') or conjugated group (Ar) etc.) are generally unreactive towards ketones but react with aldehydes to give predominantly *trans* alkenes (Figure 6.8). The geometry of the resultant double bond also depends upon other factors including the choice of solvent and the presence of lithium halides.

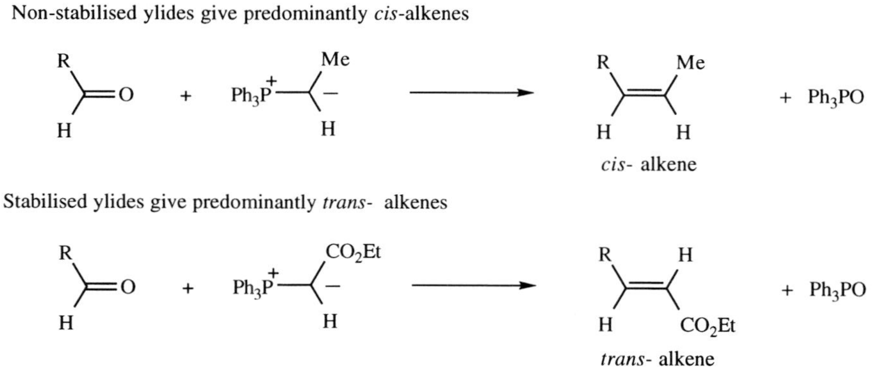

Non-stabilised ylides give predominantly *cis*-alkenes

$$
\underset{H}{\overset{R}{>}}{=}O \;+\; Ph_3\overset{+}{P}{-}\overset{Me}{\underset{H}{<}}{-} \;\longrightarrow\; \underset{H}{\overset{R}{>}}{=}\underset{H}{\overset{Me}{<}} \;+\; Ph_3PO
$$

cis- alkene

Stabilised ylides give predominantly *trans*- alkenes

$$
\underset{H}{\overset{R}{>}}{=}O \;+\; Ph_3\overset{+}{P}{-}\overset{CO_2Et}{\underset{H}{<}}{-} \;\longrightarrow\; \underset{H}{\overset{R}{>}}{=}\underset{CO_2Et}{\overset{H}{<}} \;+\; Ph_3PO
$$

trans- alkene

Figure 6.8 Stereochemistry of alkene formation

A closely related variant of the Wittig reaction employs a phosphonate ester group in place of the triphenylphosphine. This is known as the Horner–Wadsworth–Emmons reaction and has the advantage that a water–soluble by–product is produced. Many other elegant methods exist for the preparation of double bonds including the Peterson reaction and the elimination of selenoxides, sulphoxides and alkylsulphonates (see Further reading). An example of alkene formation by sulphoxide elimination in the synthesis of the natural product retronecine is given in Chapter 8.

Returning to the problem of the total synthesis of muscalure, it is evident from the above discussion that the retrosynthetic analysis of muscalure should give synthons for which the required olefin is formed either via a Wittig reaction or via an alkyne (Figure 6.9).

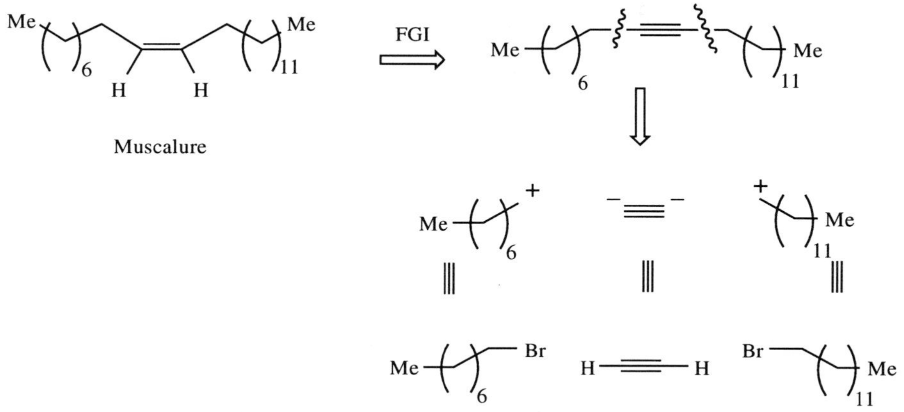

Figure 6.9 Retrosynthetic analysis of muscalure using an alkyne

Indeed both routes have been examined and the synthesis via an alkyne proved to be the better strategy. Using an alkyne (Figure 6.10), the 9,10 *cis*(Z) double bond is formed with complete regio– and stereo–control. In the case of the Wittig route (Figure 6.11), although there was complete regio–control, an 85:15 mixture of *cis*(Z) and *trans*(E) alkenes was formed which had to be separated.

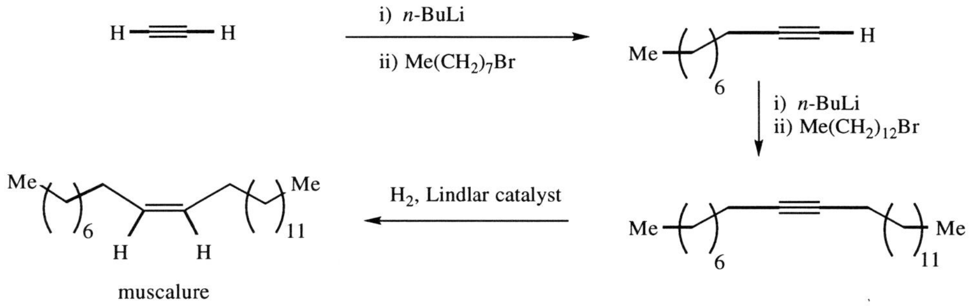

Figure 6.10 Synthesis of muscalure via an alkyne

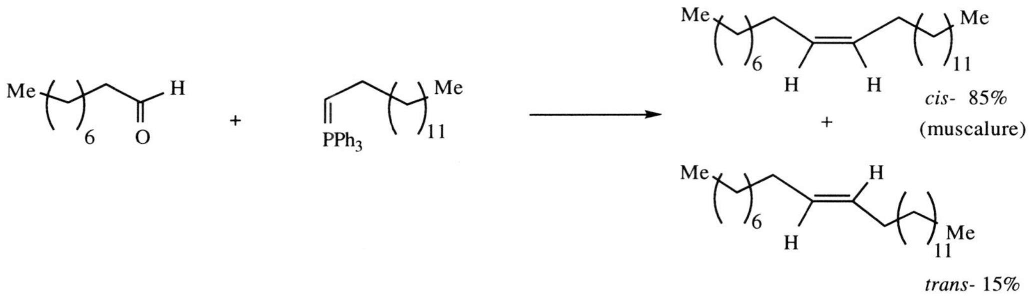

Figure 6.11   Use of a Wittig reaction for the synthesis of muscalure

6.3 Regioselective additions to alkenes

It has been illustrated above that the position of a double bond within a molecule may be controlled. Once formed, the alkene may undergo further regioselective reactions such as hydration.

Hydration

The π-bond of an alkene is electron rich and hence has a tendency to react with electrophilic reagents. If the alkene is unsymmetrical the regiochemistry of the addition reaction may be controlled. For example hydration of a double bond is usually accomplished commercially by passing the alkene into a mixture of sulphuric acid and water. With an unsymmetrical alkene the initial protonation occurs so as to afford the more stable carbocation. Since alkyl substituents stabilise carbocations, the proton adds to the less substituted carbon of the double bond (often referred to as *Markovnikov* addition, Figure 6.12).

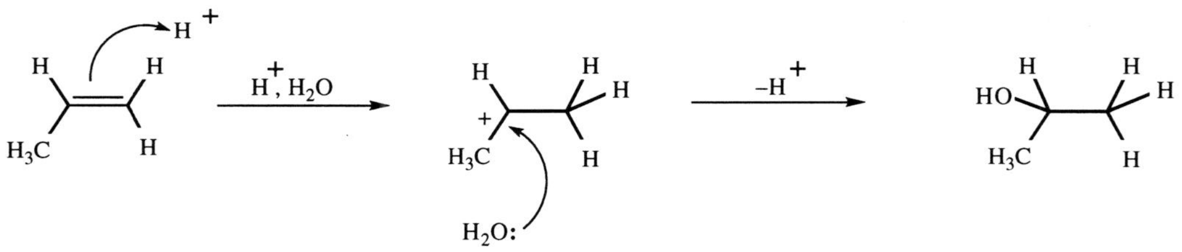

Figure 6.12   Markovnikov hydration of an alkene

Although direct hydration is an important industrial process, it is seldom used as a laboratory procedure since mixtures of products may be obtained as a result of alkyl migrations (see previous section).

Oxymercuration–demercuration

Ac is an abbreviation

for

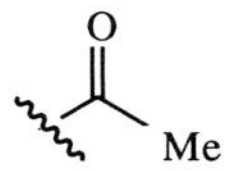

A much more reliable method for the small–scale hydration of olefins involves the use of mercuric ion, Hg(II), which results in the same regioselective overall addition of water to the double bond, i.e. the result is *Markovnikov* hydration of the double bond (Figure 6.13).

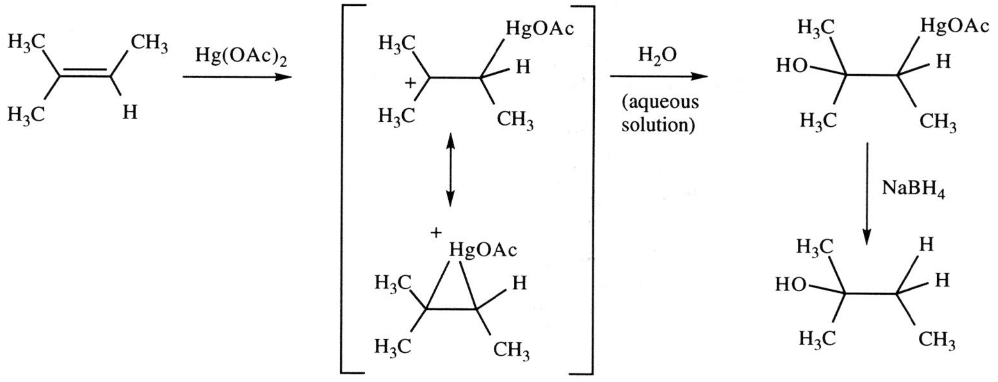

Figure 6.13 Markovnikov hydration using mercuric acetate

Hydroboration

The reaction of alkenes with diborane was discovered by Professor H. C. Brown in 1956 and it has become one of the most important reactions in the repertoire of the synthetic chemist. This was reflected by the fact that Brown was awarded a Nobel prize in 1979 for his invaluable contributions to synthetic chemistry.

In the hydroboration reaction, the boron–hydrogen bond adds rapidly to an alkene with the boron atom generally being attached to the less substituted and hence less sterically congested carbon (Figure 6.14). The alkylborane products are usually not isolated but converted to the alcohol with hydrogen peroxide and sodium hydroxide. The net result of hydroboration and oxidation–hydrolysis is *anti–Markovnikov* hydration of the double bond. For further detailed descriptions of this mechanism and indeed all of the reactions described above see Further reading.

The first step of hydroboration involves addition of borane across the double bond of the substrate, with formation of the carbon–boron bond at the less sterically hindered end

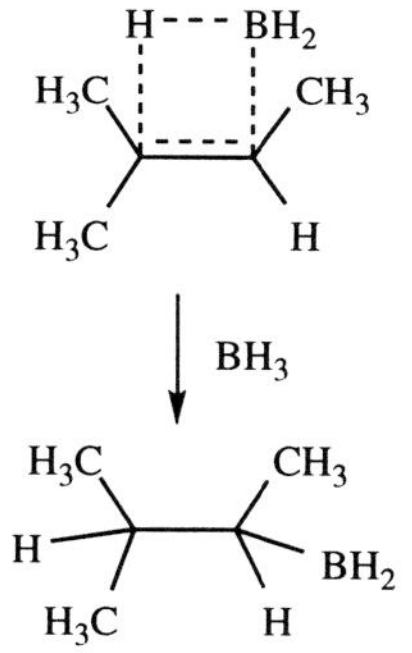

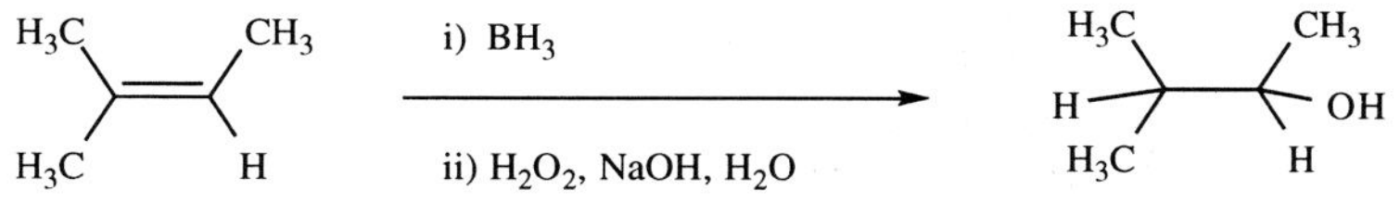

Figure 6.14 Hydroboration of an alkene

6.4 Electrophilic aromatic substitution

In contrast to the reaction of alkenes, aromatic compounds give substitution, rather than addition reactions with electrophiles. The position of substitution of the electrophile on the aromatic ring depends on the substituent(s) already present on the ring. If the aromatic compound has a substituent which will stabilise an adjacent positive charge then the electrophile adds predominantly at the *ortho–* or *para–* positions e.g. nitration of methylbenzene (toluene) giving a mixture of 2– and 4– nitrotoluene (Figure 6.15).

Other *ortho/para* directing groups:

alkyl
NHCOR
OH, OR
F, Cl, Br, I

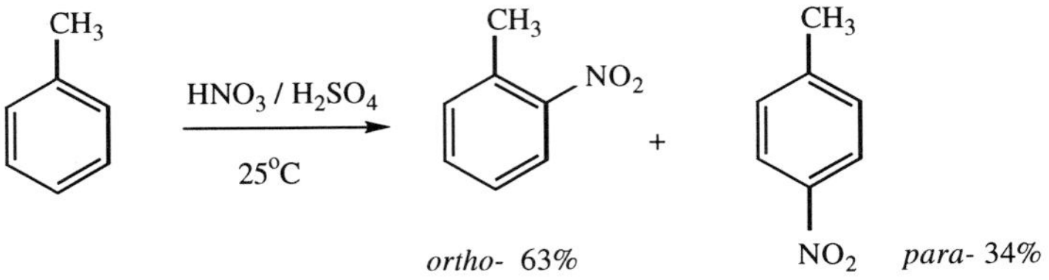

Figure 6.15   Nitration of toluene

ortho- 63%   NO$_2$ *para-* 34%

In contrast if the substituent destabilises the hexadienyl cation, regioselective attack occurs at the *meta-* position of the ring. Nitration of benzonitrile gives 3-nitrobenzonitrile as the major product (Figure 6.16). More details on aromatic substitution reactions may be found in the section on further reading at the end of this and other chapters.

Other *meta-* directing groups:

NO$_2$
CN
CHO, CO$_2$H, CO$_2$R, COR

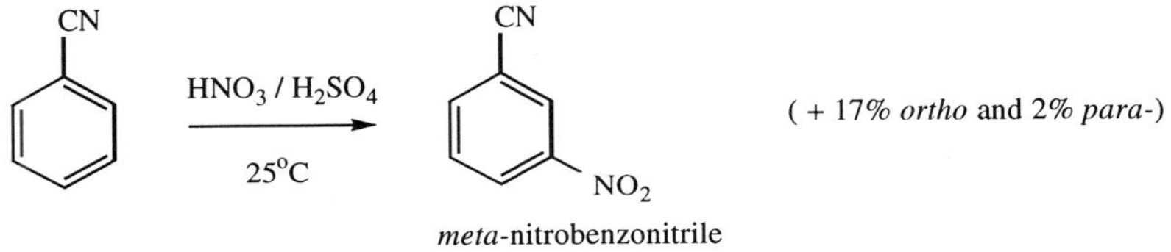

(+ 17% *ortho* and 2% *para-*)

meta-nitrobenzonitrile

Figure 6.16   Nitration of benzonitrile

6.5 Regioselective alkylation of ketones

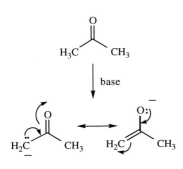

Figure 6.17 Formation of an enolate

As outlined in Chapters 2 to 4, a powerful method for the formation of carbon–carbon bonds involves the alkylation of carbonyl compounds, i.e. the abstraction of the acidic proton α to a carbonyl group with base followed by reaction of the resultant enolate with an electrophile such as an aldehyde or alkyl halide. The enolate is stabilised by resonance and the negative charge resides partially on the oxygen and partially on the carbon atom (Figure 6.17). Both forms are important in the nucleophilic reactions of enolate ions and electrophiles may add at carbon or oxygen. A species which may react in this manner to give two products is known as *ambident* (from the Latin *ambi* –both, and *dens*–tooth, hence 'two fanged').

Alkylation of the enolate from propanone with alkyl halides and carbonyl compounds generally results in reaction at carbon (C-alkylation). In fact *most reactions of enolates with electrophiles occur on carbon*. An important exception is the reaction with trialkylsilyl chlorides which takes place at the oxygen atom (O–silylation) to give silyl enol ethers containing strong oxygen–silicon bonds (Figure 6.18). Silyl enol ethers are important intermediates in organic synthesis and are described in more detail below.

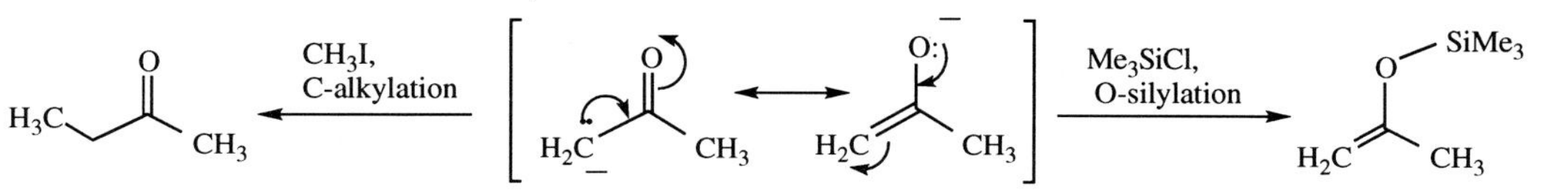

Figure 6.18 Reactions of enolates with electrophiles

In the case of an unsymmetrical ketone there is a problem in controlling the position of alkylation. For example, abstraction of an acidic proton from butanone potentially gives rise to two isomeric enolate anions (Figure 6.19). Deprotonation is more rapid at the less substituted, and therefore the less hindered, carbon resulting in formation of the kinetic enolate (i.e. the fastest to form). However deprotonation at the more substituted carbon gives the more substituted thermodynamic enolate (i.e. the more stable). The reaction of each enolate with an electrophile such as benzyl bromide will give a different product.

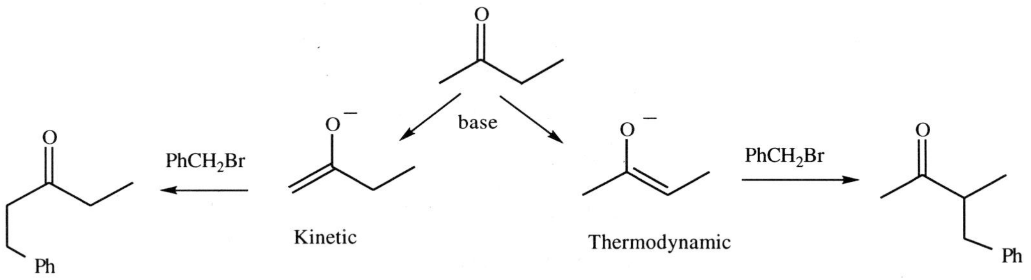

Figure 6.19 Alkylation of thermodynamic and kinetic enolates

How might the regioselective alkylation of an unsymmetrical ketone be controlled ?

There are two basic strategies:

i. Selective formation of kinetic or thermodynamic enolates and trimethylsilyl enol ethers:

By altering the reaction conditions used for the preparation of the enolate ion, it is possible to produce selectively the thermodynamic or kinetic enolate. We can measure the ratio between the enolates by adding trimethylsilyl chloride to the reaction mixture which traps the enolates as the trimethylsilyl enol ether derivatives (Figure 6.20).

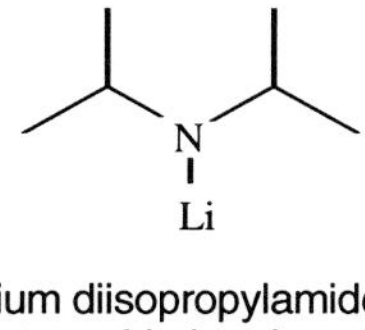

Ideal conditions for the formation of kinetic enolates involve the use of a strong hindered base (such as lithium diisopropylamide) at low temperature (ca. −78°C) in a non-polar solvent (e.g. THF or ether). These conditions favour an *irreversible* deprotonation of the ketone, an essential requirement for kinetic enolate formation. In contrast, heating the carbonyl compound to 130°C with a weak base (triethylamine) in the presence of trimethylsilyl

Lithium diisopropylamide (LDA) is a strong hindered organic base.

chloride gives the trimethylsilyl enol ether of the thermodynamic enolate. In this case the product is formed via a *reversible* enolisation process.

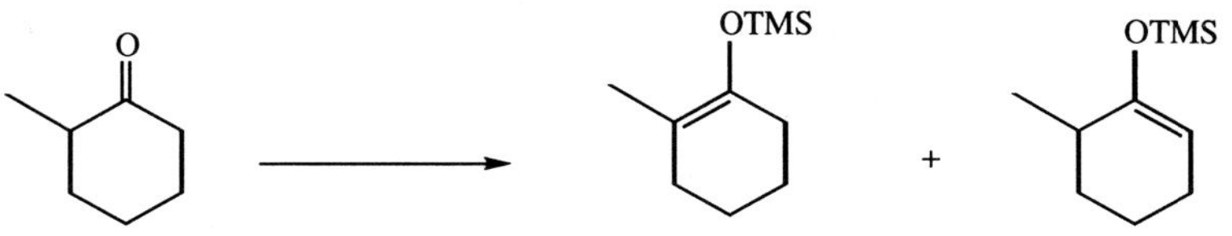

	Thermodynamic product		Kinetic product
LDA, -78°C, THF, Me₃SiCl	1	:	99
Et₃N, 130°C, Me₃SiCl	12	:	1

Figure 6.20 Kinetic and thermodynamic deprotonation of 2-methylcyclohexanone

The enol ethers may be isolated and used in further reactions. Separation of the enol ethers followed by reaction with methyllithium gives the free lithium enolate which may be subsequently reacted with an alkyl halide or other electrophile (Figure 6.21). Alternatively a powerful electrophile may be capable of reaction directly with the double bond of the silyl ether in which case silyl removal is the second step. A Lewis acid may be employed to promote this reaction.

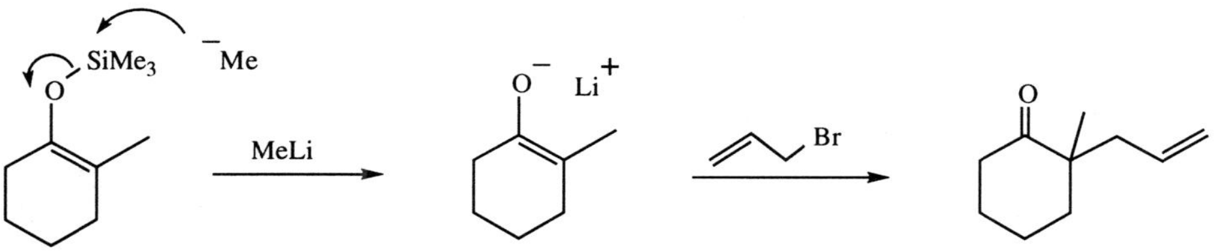

Figure 6.21 Regioselective reactions of silyl enol ethers

ii. *Addition of an activating group such that the acidity of the protons on either side of the carbonyl group may be differentiated*

Monoalkylation of an unsymmetrical ketone is difficult to control if the protons on either side of the carbonyl group are of similar acidity. An excellent way to control the deprotonation is to add an activating group to the substrate such that protons at one position are significantly more reactive than at the other. The most commonly used group is an ester function β to the carbonyl group to be alkylated. Not only does the ester serve to increase the acidity of the protons α to both carbonyl functions, it may also be simply removed at the end of the synthesis by hydrolysis and decarboxylation. An example of how the alkylation of such a β–ketoester may be used in the regiocontrolled dialkylation of a ketone is shown in Figure 6.22.

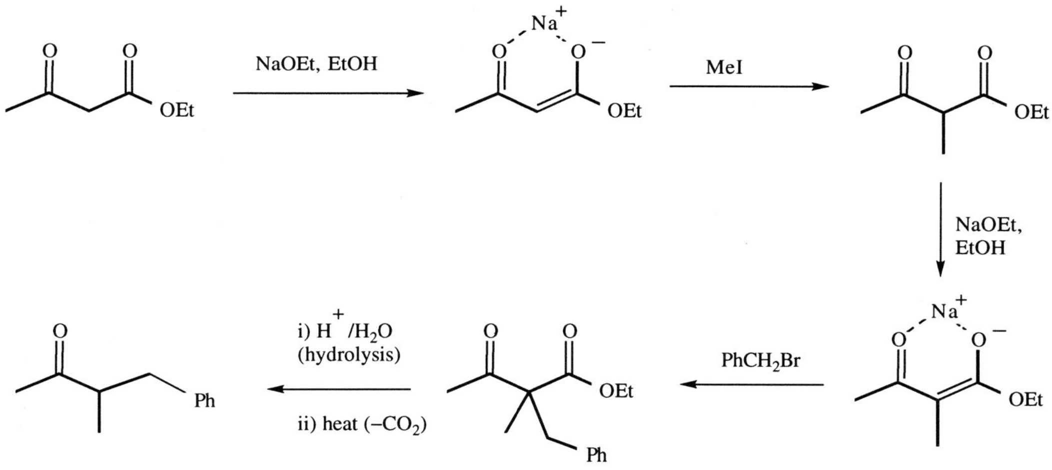

Figure 6.22 The use of an activating group to direct alkylation reactions

The mechanism of the decarboxylation step involves elimination of carbon dioxide from the β–keto acid formed by hydrolysis of the ester. The initial product is an enol which rapidly reverts to the ketone structure (Figure 6.23). Two reagents commonly used in synthesis which benefit from this 'double–activation effect' are ethyl acetoacetate ($CH_3COCH_2CO_2Et$) and diethyl malonate ($EtO_2CCH_2CO_2Et$). These reagents are effectively equivalent to the synthons $CH_3COCH_2^-$ and $HO_2CCH_2^-$ respectively (see Table 2.1, page 13).

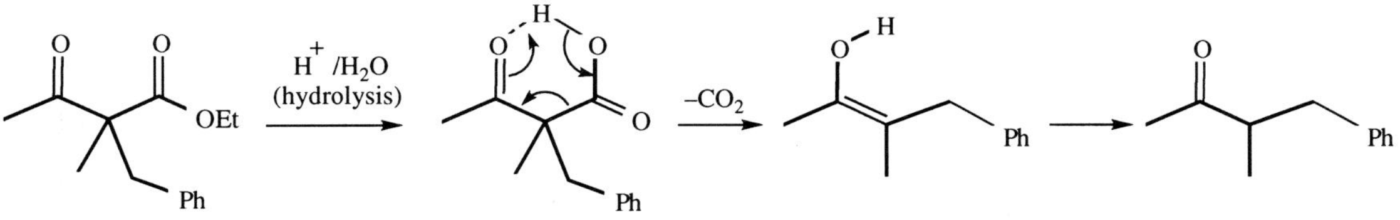

Figure 6.23 Mechanism of decarboxylation of β–keto esters

6.6 Regioselective addition of nucleophiles to α,β–unsaturated carbonyl compounds

The addition of nucleophiles to α,β–unsaturated carbonyl compounds is another example of a class of reaction which may have more than one regiochemical outcome. In an α,β–unsaturated carbonyl compound the functional group is polarised such that there are two potential sites for attack by nucleophiles (see page 3). Addition directly to the carbonyl carbon is known as 1,2–addition whereas attack at the alkene carbon is 1,4– or conjugate addition (Figure 6.24).

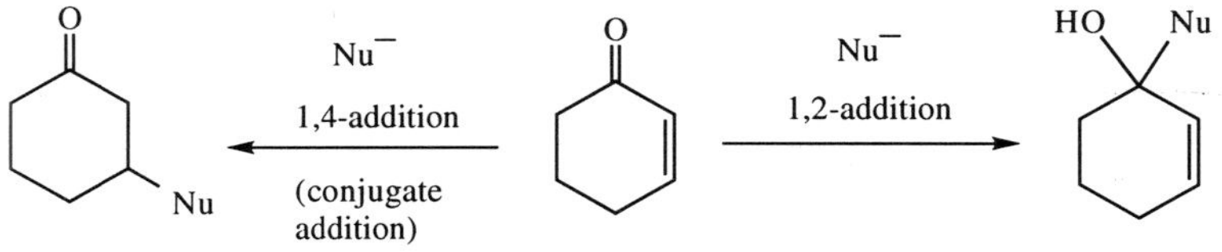

Figure 6.24   Regioselective additions to enones

As a general rule strongly basic nucleophiles tend to attach directly to the carbonyl carbon (i.e. 1,2–addition), whereas weakly basic nucleophiles tend to undergo 1,4–addition. A list of selectivities is given in Table 6.1. If conjugate addition involves attack by a carbon nucleophile, this is known as the Michael reaction.

Nucleophilic atom	Nucleophile	1,4-Addition product	1,2-Addition product
Carbon nucleophiles	RLi	✗	✓
	R_2CuLi	✓	✗
	RMgBr	✗	✓
	RMgBr/CuI	✓	✗
	$NaCH (CO_2Et)_2$	✓	✗
Hydrogen nucleophiles	$LiAlH_4$	✗	✓
	$NaBH_4$/CuI	✓	✗
	$NaBH_4$/$CeCl_3$	✗	✓
Heteroatom nucleophiles	RNH_2	✓	✗
	RSNa	✓	✗
	RONa	✓	✗

Table 6.1 Predominant product of addition of nucleophiles to α,β–unsaturated carbonyl compounds.

Nucleophilic attack on a α,β-unsaturated carbonyl compound has tremendous scope in synthesis because the resultant enolate from a conjugate addition reaction may be trapped by an electrophile as shown in Figure 6.25. The stereochemical outcome of these reactions is discussed in Chapter 7.

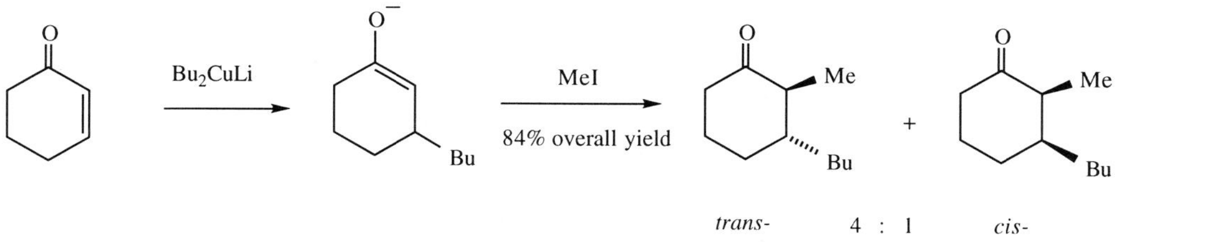

Figure 6.25 Conjugate addition followed by addition of an electrophile

6.7 Regioselective addition of nucleophiles to epoxides

Epoxides (oxiranes) are prone to attack by a wide range of nucleophiles under both acidic and basic conditions resulting in ring–opening. In the case of an unsymmetrical epoxide the site of attack depends largely on the reaction conditions. Consider for example the reaction of propylene oxide with sodium methoxide (Figure 6.26). There are two possible sites of attack: a) at C-1 giving 1–methoxy–2–propanol or b) at C-2 giving 2–methoxy–1–propanol. With sodium methoxide an S_N2 attack occurs regioselectively at the less sterically hindered position, i.e. at the less substituted carbon atom. The stereochemical outcome of the reaction is discussed in Chapter 7.

Reaction of propylene oxide under acid-catalysed conditions may also potentially give two products. However this reaction proceeds *via* a different mechanism to the base catalysed reaction. Initial protonation of the oxirane oxygen gives the oxonium ion intermediate with substantially polarised carbon–oxygen bonds. The partial positive charge is more stabilised on the secondary carbon atom than the primary. This uneven charge distribution counteracts the steric effects, hence reaction of propylene oxide with HCl/MeOH gives the primary alcohol as the major product (Figure 6.27).

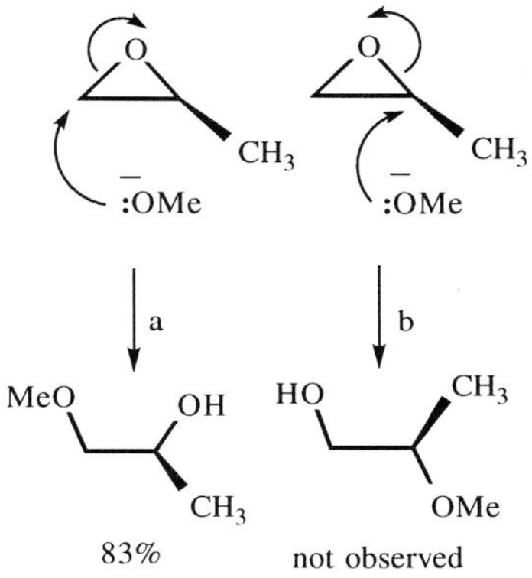

Figure 6.26 Epoxide opening by a nucleophile

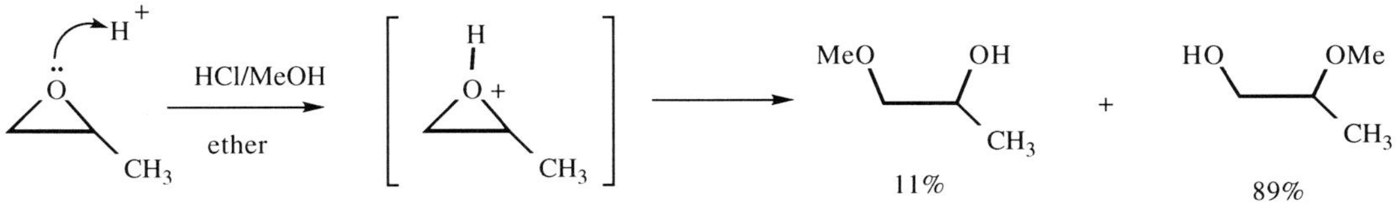

Figure 6.27 Acid-catalysed nucleophilic attack on an epoxide

6.8 Regioselective oxidation of ketones to esters – the Baeyer-Villiger reaction

A further example of a regioselective reaction is the Baeyer–Villiger oxidation which transforms acyclic ketones to esters and cyclic ketones to lactones. The reaction proceeds by the mechanism shown below involving a 1,2–migration of an alkyl group to the electron deficient oxygen atom (Figure 6.28).

Baeyer-Villiger oxidation:

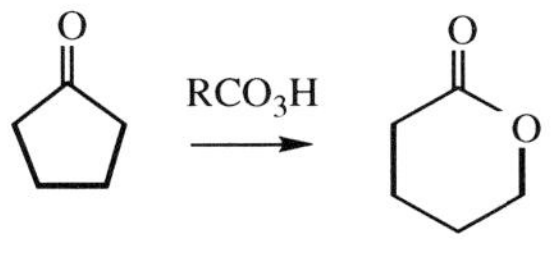

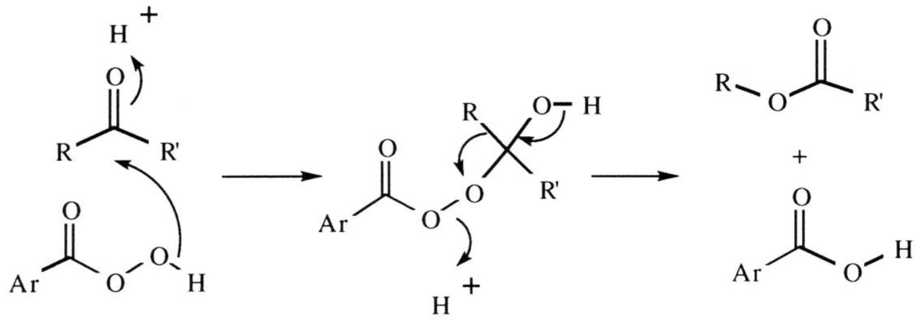

Figure 6.28 Mechanism of the Baeyer–Villiger reaction

The overall result of the Baeyer–Villiger oxidation is insertion of an oxygen atom into one of the bonds to the carbonyl group. Unsymmetrical ketones can potentially give two oxidation products. However there is sufficient difference in the migratory aptitude of alkyl groups for the reaction to proceed regioselectively (Figure 6.29). As a result oxygen is inserted between the carbonyl carbon and the more substituted α–carbon.

Migratory aptitude: hydrogen > tertiary alkyl > secondary alkyl > phenyl > primary alkyl >methyl

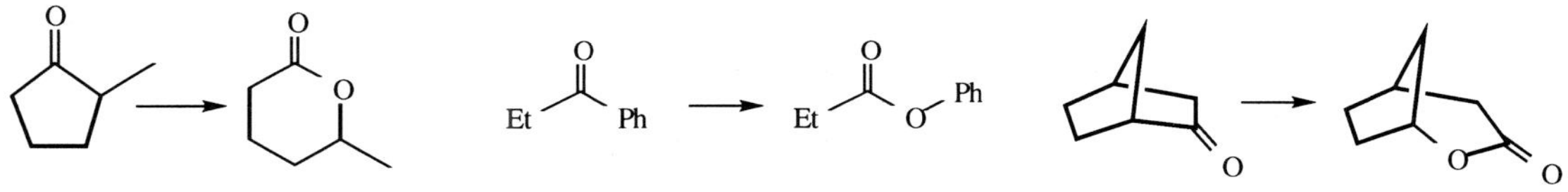

Figure 6.29 Regioselective Baeyer–Villiger oxidations

The Baeyer–Villiger reaction has been used with total regiocontrol in the preparation of Grieco lactone (Figure 6.30). This material is a valuable starting material for the synthesis of *cis*—jasmone, used in the perfumery industry. The reaction has also been used in prostaglandin synthesis, the biological significance of which was outlined in Chapter 1.

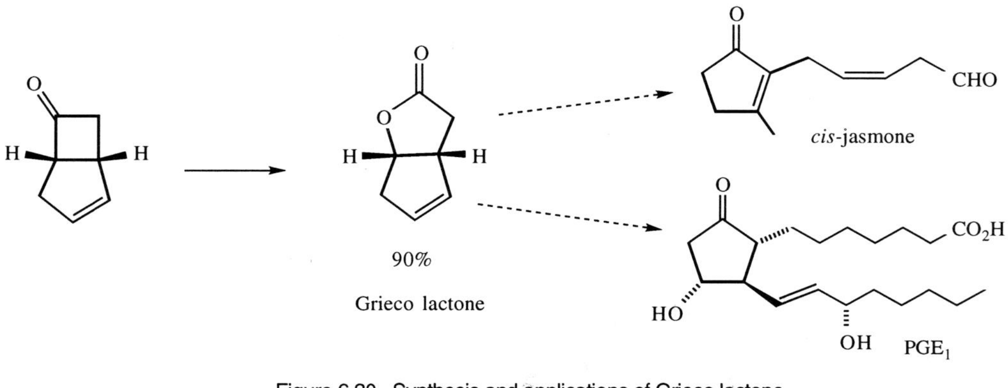

90%

Grieco lactone

cis-jasmone

PGE₁

Figure 6.30 Synthesis and applications of Grieco lactone

6.9 Practice examples

Bearing in mind the need to control regiochemistry, devise syntheses of the target molecules below from the starting materials indicated:

Further reading

S. E. Thomas. *Organic Synthesis: The Role of Boron and Silicon,* (Oxford Chemistry Primer no. 1), Oxford University Press (1991).

P. Sykes. *A Guidebook to Mechanism in Organic Chemistry* (6th edn), Longman (1986).

L. Harwood. *Polar Rearrangements* (Oxford Chemistry Primer no. 5), Oxford University Press (1992).

R. S. Ward. *Bifunctional Compounds* (Oxford Chemistry Primer no.17), Oxford University Press (1994).

7 Selectivity III: Stereoselectivity

7.1 Introduction

The biological and physical properties of organic molecules used as drugs, insecticides, plant growth regulators, perfumes, etc., depend to a large extent on the *stereochemistry* of substituents and functional groups. Stereochemistry also has an important effect on the reactivity of compounds. For example, the oxidation of *cis*–4–*t*–butylcyclohexanol occurs at a rate three times as fast as that of the *trans*– isomer using chromium trioxide in acetic acid.

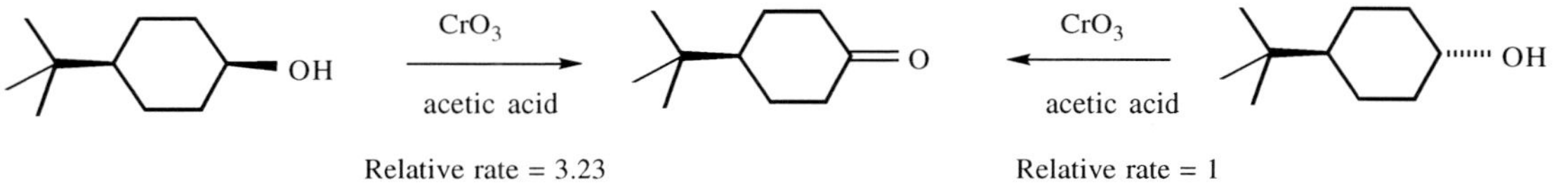

Relative rate = 3.23   Relative rate = 1

Figure 7.1   Relative rate of oxidation of 4–*t*–butylcyclohexanol stereoisomers

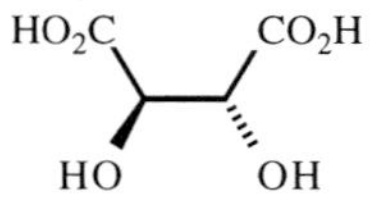

(+)–*R,R*–tartaric acid

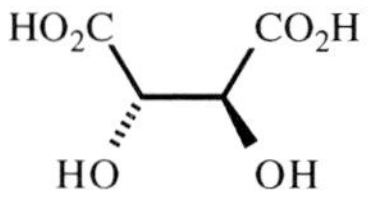

(−)–*S,S*–tartaric acid

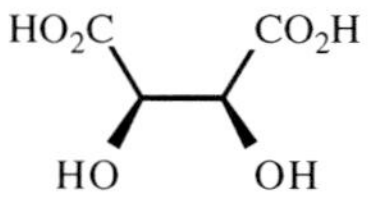

meso–tartaric acid

Stereoisomers have the same carbon framework and the substituents have identical regiochemistry; however the isomers differ in their three-dimensional spatial arrangement of atoms within the molecule. Each isomer has a unique *relative configuration* that can only be converted to another by a process of breaking and making bonds.

Stereoisomers containing two or more stereogenic centres are described as being diastereomers (or *diastereoisomers*). Diastereomers are stereoisomers that are non–enantiomers, i.e. not mirror images of each other. A compound with n stereocentres will have a *maximum* of 2^n stereoisomers. However the symmetrical compound tartaric acid may exist in only three forms instead of the theoretical maximum of four. Two of these, the (+)*R,R*– and (-)*S,S*– forms, are enantiomers of each other, and therefore have opposite rotation signs. The third form (i.e. the *R,S* and *S,R* diastereomers) has no optical activity and is known as *meso*–tartaric acid.

7.2 Stereospecific reactions

Some chemical reactions have a mechanism which demands a specific stereochemical outcome– these are known as *stereospecific reactions*, e.g. S_N2 reactions. An S_N2 reaction involves a concerted displacement of a good leaving group by a nucleophile via a stereospecific backside attack resulting in an inversion of configuration at a stereogenic centre.

1,2–Diols may be prepared from alkenes using stereospecific oxidative procedures. Reaction of cyclohexene with osmium tetroxide gives the osmate ester which may be cleaved to give the product with the two hydroxyl groups on the same side of the molecule, the *syn* or *cis* diol (Figure 7.2). In contrast, reaction of cyclohexene with a peracid gives an epoxide which, on treatment with aqueous acid, gives attack from the backside of the epoxide yielding the product with the two hydroxyl groups on different sides of the molecule (the *anti* or *trans* diol).

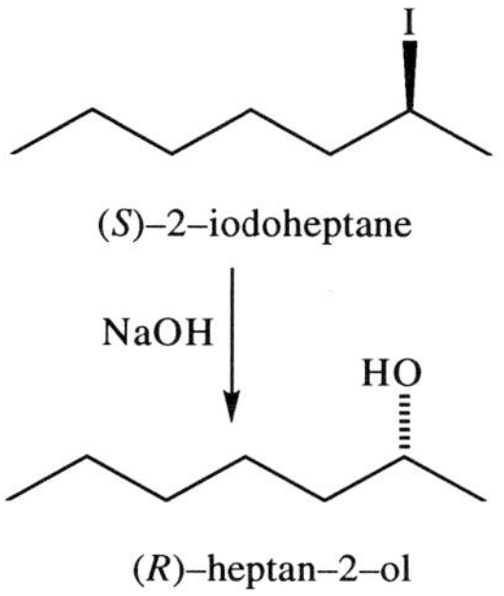

(*S*)–2–iodoheptane

NaOH

(*R*)–heptan–2–ol

Stereospecific S_N2 reaction

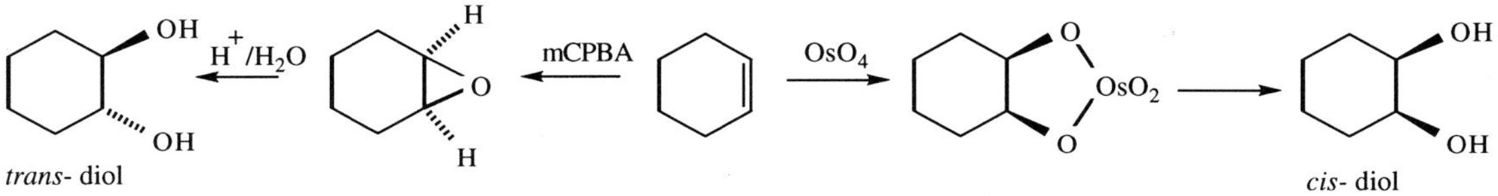

trans- diol

cis- diol

Figure 7.2 Stereospecific synthesis of 1,2–diols

Another method for the preparation of 1,2–diols from alkenes requires the use of silver acetate and iodine (Figure 7.3). The stereochemical outcome of the reaction depends critically on whether or not water is present, and either the *cis* or the *trans* diol may be prepared selectively in good yield. If water is present then a *cis*– diol is formed (Woodward reaction), if not then a *trans*– diol is formed after hydrolysis of the diacetate (Prevost reaction).

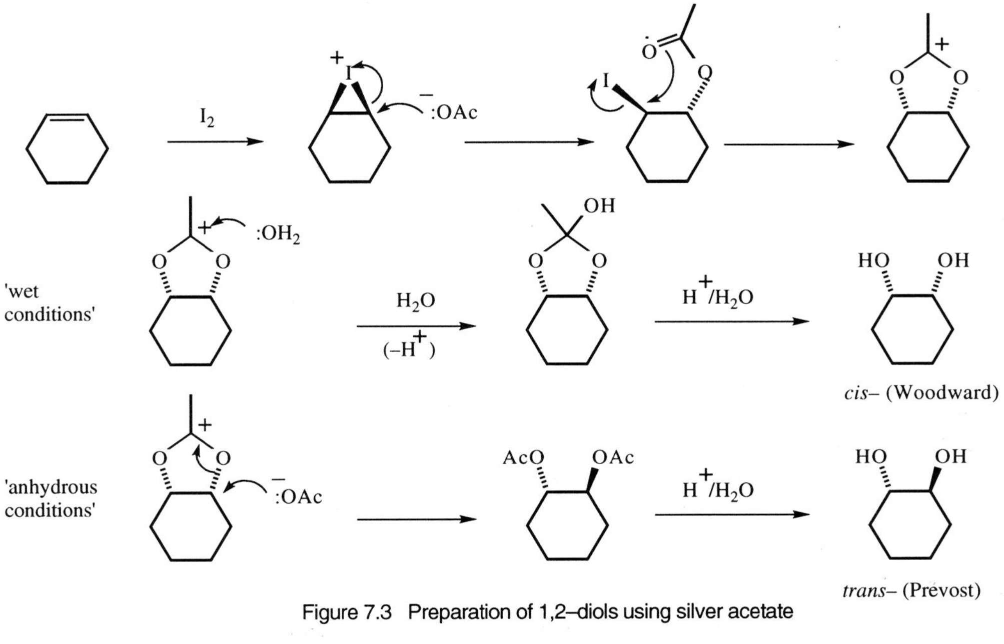

'wet conditions'

'anhydrous conditions'

cis– (Woodward)

trans– (Prevost)

Figure 7.3 Preparation of 1,2–diols using silver acetate

7.3 Stereoselective reactions

Stereoselective reactions are reactions whose mechanisms offer alternative pathways so that the reaction may proceed either via the most favourable pathway (kinetic control) or via the pathway which gives the most stable stereoisomer as the major product (thermodynamic control). Most commonly, selectivity of this type is achieved through the presence of a physical barrier which hinders the formation of one isomer (steric hindrance). In general it is simpler to achieve this type of control in rigid cyclic systems.

Cyclic molecules

An example of a stereoselective reaction in a cyclic molecule is the reaction of 4–methylcyclohexenone with lithium dimethylcuprate which results in conjugate addition of a methyl group to the β–position of the enone (i.e. a regioselective reaction, Figure 7.4). However two stereoisomers may be formed, one in which the newly introduced methyl group is *cis* to the existing methyl group and the other isomer in which the methyl groups are *trans*. In practice the *trans*– isomer is the major product since approach of the bulky cuprate reagent occurs predominantly on the less sterically hindered face of the enone, i.e. away from the 4–methyl group.

A further substituent may be introduced α to the carbonyl group if the enolate which is formed in the conjugate addition reaction is alkylated by an electrophile. Steric factors once more control the stereochemical outcome of this reaction and the preferred product is the one in which the substituents α and β to the carbonyl group are *trans* to each other. Sequential control of stereochemistry in this way has been used in the synthesis of prostaglandins, as illustrated in Figure 7.5.

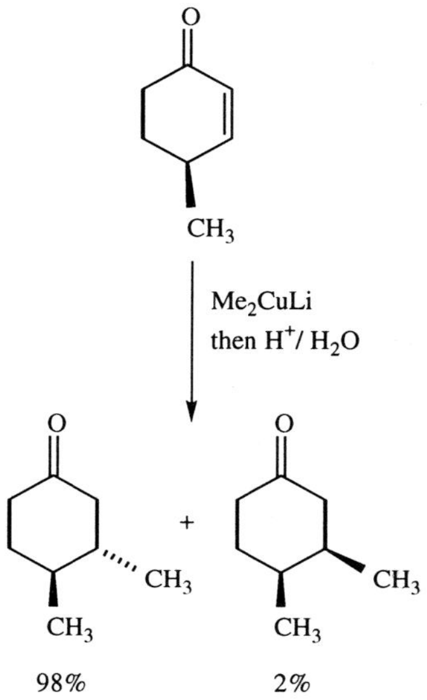

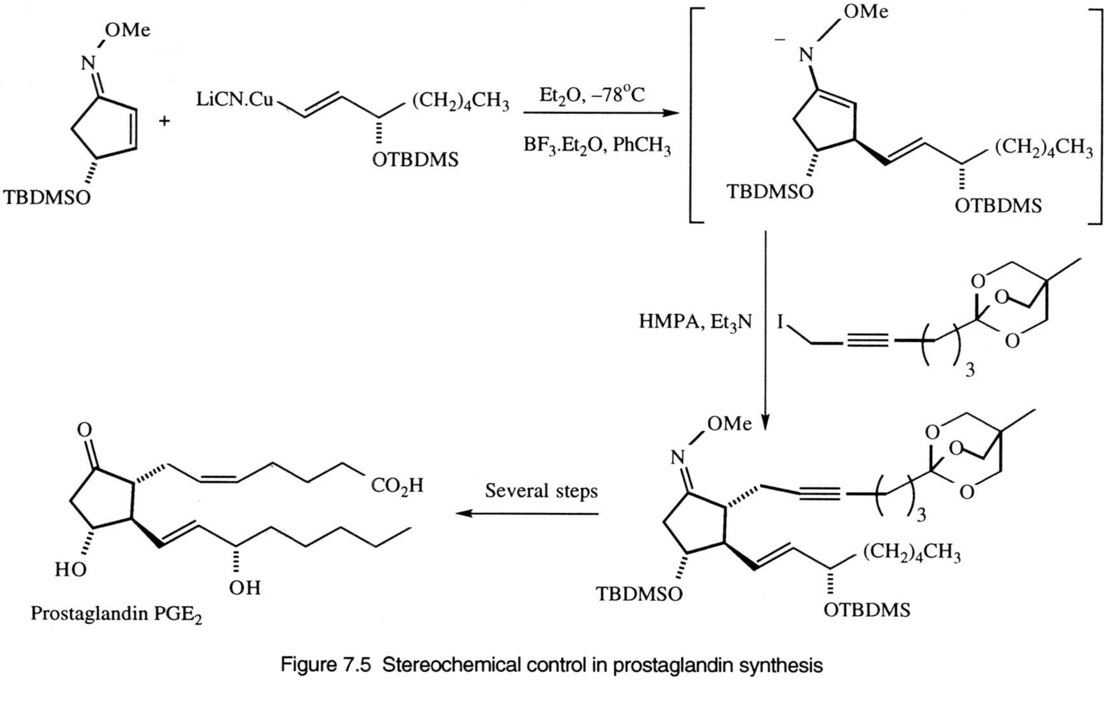

Figure 7.4 Stereoselective conjugate addition reaction

Figure 7.5 Stereochemical control in prostaglandin synthesis

Acyclic molecules

Stereocontrol in acyclic molecules is rather more difficult to achieve than in cyclic cases because of the greater flexibility of the compounds. However one case in which we *can* predict the stereochemical outcome is in the addition of a nucleophile to an acyclic ketone. As we have already seen, addition of a nucleophile to a carbonyl compound may occur under acidic or basic conditions. The addition of a nucleophile to a simple unsymmetrical ketone gives a tertiary alcohol. The product will always be a racemate (i.e. 1:1 mixture of enantiomers) because the initial attack by the nucleophile will statistically be equally likely from above or below the plane of the molecule (Figure 7.6).

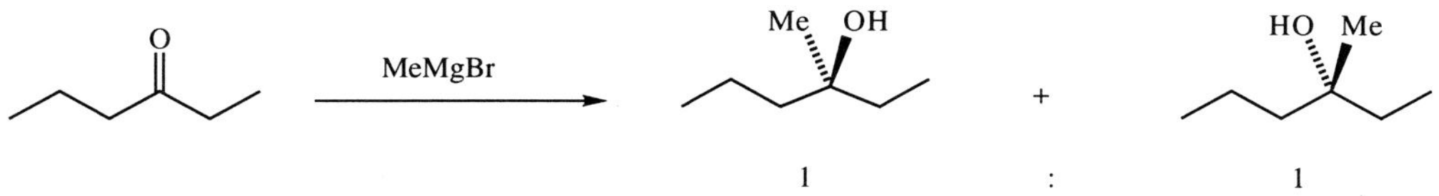

Figure 7.6 Addition of nucleophiles to ketones

However if the carbon atom α to the carbonyl group bears a stereogenic centre, then the two faces of the carbonyl group are no longer equivalent. Addition of the nucleophile from above or below the plane of the carbonyl group is no longer statistically equal and a mixture of diastereoisomers is formed (Figure 7.7).

When the addition of the nucleophile is reversible, it is likely that the thermodynamically more stable of the two possible products will be formed predominantly. In contrast for essentially irreversible reactions, e.g. a Grignard reaction or reduction, the kinetic product is likely to be the major diastereomer. The two faces of the carbonyl group are described as *re* or *si*; and the products are described as *syn* and *anti*. These terms are defined in the Glossary.

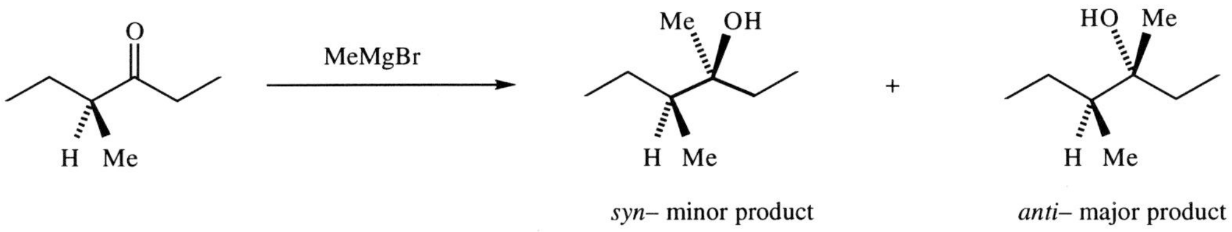

Addition of a nucleophile to a carbonyl compound is irreversible if the nucleophile is hydride (H^-) or a carbanion (R^-), but generally is reversible if the nucleophile is an alkoxide (RO^-), halide (X^-) or an amine (RNH_2).

Figure 7.7 Diastereoselective addition to a ketone

Several models have been formulated to rationalise the stereochemical outcome of these reactions. Although the earliest model, that of *Cram*, is often cited, the *Felkin–Ahn* model is now preferred. The basis of this model is that the face of the carbonyl group to which addition of the nucleophile is preferred is controlled by the positions in space of the substituents on the stereogenic centre relative to the carbonyl group. The most reactive conformers are assumed to be those in which the bond to the largest group on the stereogenic centre is perpendicular to the plane of the carbonyl group. The two remaining substituents on the stereogenic centre may then be arranged in two ways, i.e. with the smaller group towards or away from the carbonyl oxygen, (**1**) or (**2**) respectively.

Remember that a nucleophile adds to a carbonyl group at an optimum angle of 107°. Comparison of the two possible reactive conformers reveals the nucleophile will be hindered by the medium sized group in (**1**) but only by the small group in (**2**). Hence (**4**) is the major product (Figure 7.8).

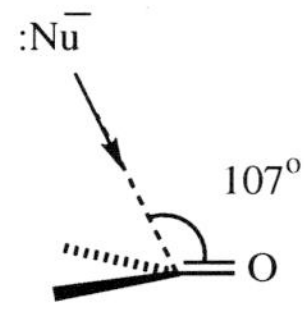

Optimum angle of nucleophilic attack on a carbonyl group

Application of the Felkin-Ahn model:

1) Identify the largest and the smallest groups

2) Place largest group perpendicular to carbonyl in Newman projection.

3) Nucleophile will prefer to attack the face of the carbonyl group depicted in (**2**) to give (**4**) as major product.

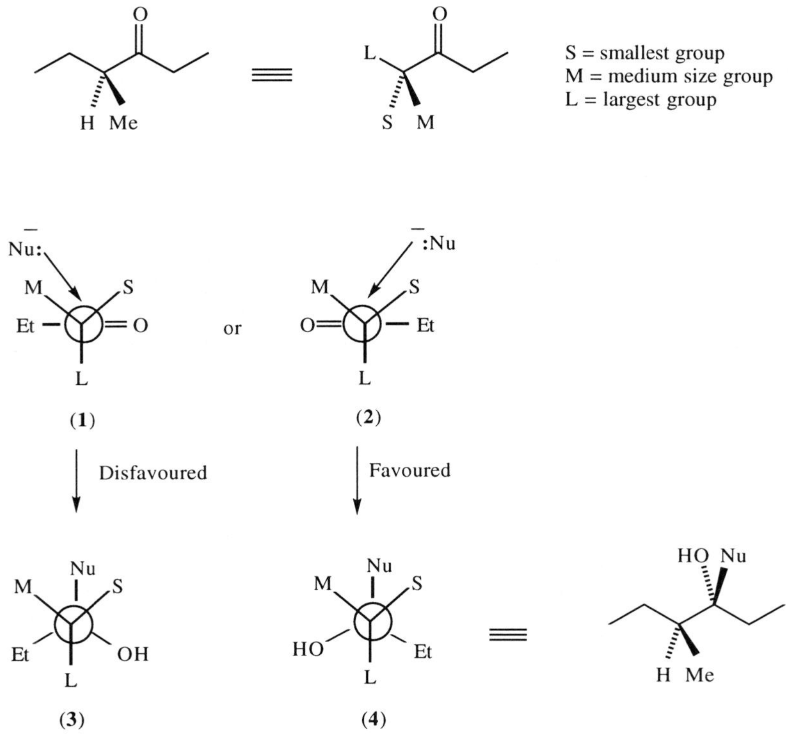

S = smallest group
M = medium size group
L = largest group

Figure 7.8 Application of the Felkin–Ahn model

The stereochemical outcome of the reaction may be different when there is a coordinating group, such as methoxyl, at the centre α– to the ketone. In this case chelation of the counterion (i.e. the magnesium cation in the case of a Grignard reagent) takes place which effectively locks the molecule in a cyclic form. The large group then blocks approach of a nucleophile to one face and addition takes place predominantly from the opposite side (Figure 7.9).

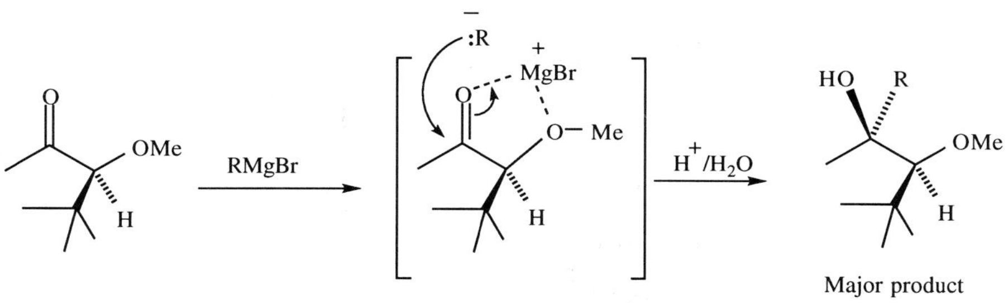

Figure 7.9 Chelation controlled nucleophilic addition to a ketone

The guidelines given above provide you with an introduction to how stereocontrol can be achieved in synthetic manipulations of both cyclic and acyclic molecules. Further details may be found in the references cited at the end of this and previous chapters.

7.4 Conclusions

The discussion presented in Chapters 5, 6 and 7 should provide all of the most essential basic knowledge that you require to devise synthetic approaches to target molecules in cases where more than one isomer may be formed. Below are some examples of target molecules for which you may practice designing syntheses.

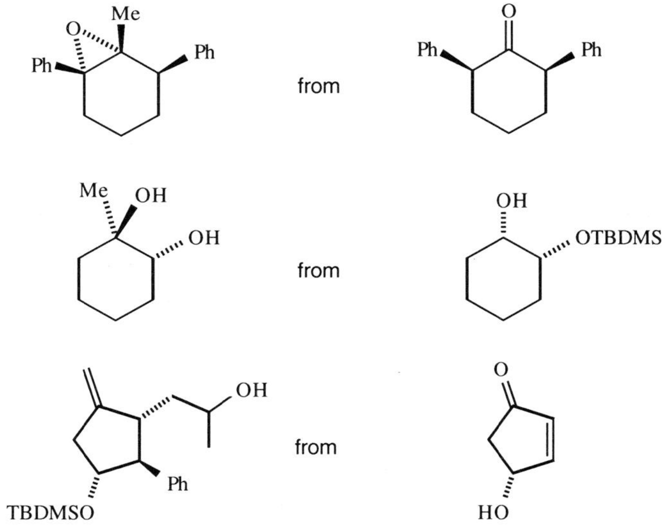

from

from

from

Further reading

E. L. Eliel, S. H. Wilen and L. N. Mander. *Stereochemistry of Organic Compounds*, Wiley (1994).

A. Bassindale. *The Third Dimension in Organic Synthesis*, Wiley (1984).

M. Nogradi. *Stereoselective Synthesis,* VCH (1987).

8   Selected organic syntheses

8.1 Introduction

In the previous chapters, guidelines have been given for the logical design of synthetic routes to a particular compound from commercially available materials using retrosynthetic analysis. The importance of a good basic knowledge of potential selective functional group interconversions has also been highlighted. Further details of specific reagents and reaction conditions may be obtained from the literature. In this chapter we bring together these aspects of synthesis by considering a case history of selected syntheses of a class of natural products known as the pyrrolizidine alkaloids. We have deliberately chosen examples which introduce very little 'new chemistry' that has not already been covered in the first and second year of a degree course. With practice, you should find that you have the knowledge to devise equally viable routes to organic molecules.

8.2 Pyrrolizidine alkaloids

Background

The pyrrolizidines are a group of alkaloids (naturally occurring nitrogen containing compounds) which are characterised by the presence of the basic necine skeleton (1). The pyrrolizidine alkaloids exhibit remarkably diverse types of biological activity and have been reported to act as antitumor, hypotensive, anti–inflammatory, carcinogenic or hepatoxic agents. Indeed some of these compounds are now in clinical trials as antitumor agents.

The pyrrolizidine alkaloids often occur naturally as esters or macrocyclic bislactones (e.g. monocrotaline). The basic azabicyclo[3.3.0]octane framework (1) may have a number of substitution patterns and oxidation levels; however the most commonly found are the diols, either saturated e.g. platynecine (2), or unsaturated e.g. retronecine (3) and heliotridine (4).

(1)

Monocrotaline

Platynecine (2)

Retronecine (3)

Heliotridine (4)

The pyrrolizidine alkaloids have attracted the interest of the scientific community not only because of their remarkable biological activities but also because their diverse structures have been a stimulating challenge to the skills of the synthetic chemist (there are over 200 members of this class of compound). Many successful routes for the syntheses of these molecules have been described in the literature and we have selected four routes to compare and contrast.

In each of the four retrosynthetic analyses of the pyrrolizidine alkaloids described in this chapter, a different initial disconnection is used for the eventual assembly of the basic azabicyclo[3.3.0]octane framework. The logic of these will unfold in the discussion. The disconnections may be represented as:

Recall that in designing a synthetic route to the target molecules, it is necessary to :-

i. Construct the basic bicyclic framework

ii. Place the functional groups, i.e. the primary and secondary alcohols and in the case of heliotridine and retronecine, the alkene, in their correct positions (regiocontrol)

iii. Control the stereochemistry of the substituents.

Disconnection A

Disconnection B

Disconnection C

Disconnection D

8.3 Platynecine (2)

Retrosynthetic analysis

As a first step in the analysis of platynecine (2), add the latent polarities originating from the secondary alcohol and the nitrogen and, encouragingly, we can see immediately that they form a consonant pattern. Therefore it

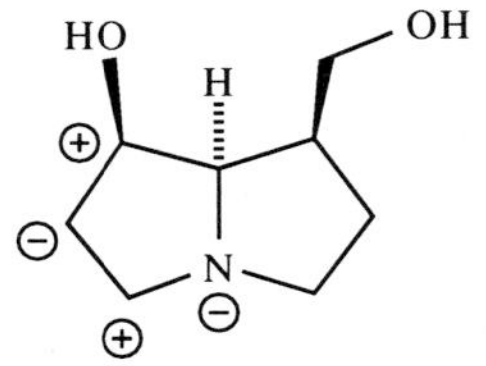

Latent polarities in (2)

(6)

(7)

would be reasonable to make disconnection 1 to give synthon (5) (Figure 8.1).

When considering possible functional groups corresponding to the negatively charged part of the synthon (5) it is apparent that the anion may be either in the form of a Grignard reagent or generated by abstraction of an acidic proton by base. This latter method requires the use of a functional group which may not only serve to stabilise the negative charge formed on abstraction of the acidic proton, but also removed once the appropriate carbon–carbon bond forming reaction has been achieved.

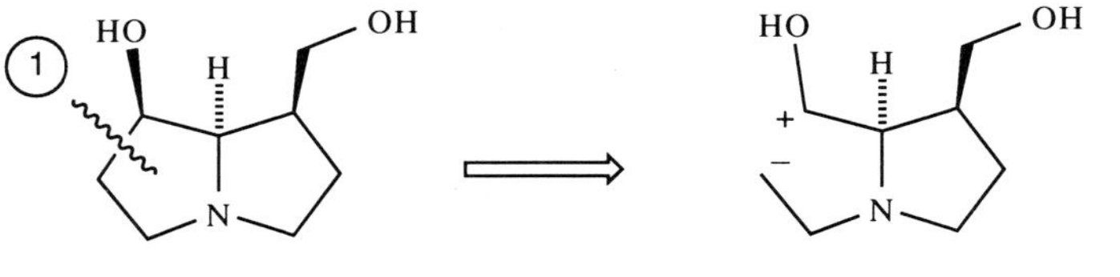

Figure 8.1

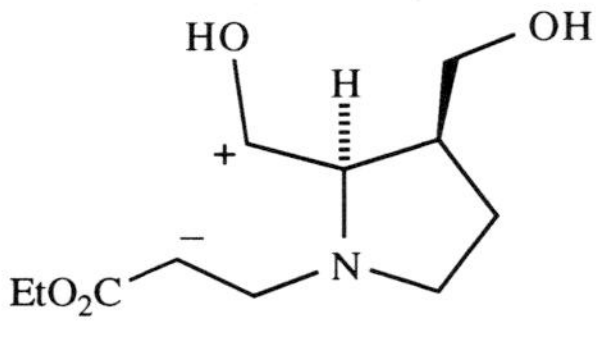

This approach was favoured by Roder, Bourauel and Wiedenfeld[1] who used an ester as the activating functional group as in synthon (6) (remember that an ester of a β-keto-ester group may be simply removed by hydrolysis and decarboxylation, see Chapter 6). Therefore the positively charged part of synthon (5) should ideally leave a ketone to facilitate this decarboxylation step after carbon–carbon bond formation has been achieved, i.e. it should also be an ester as in (7).

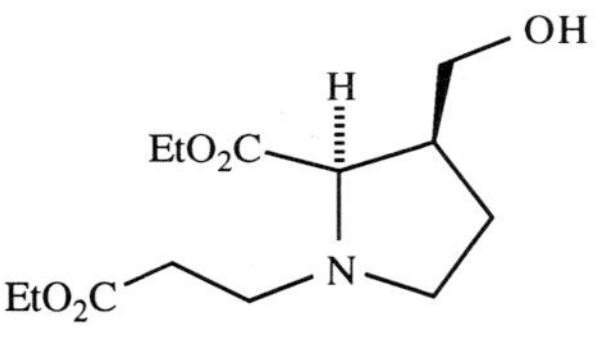

Therefore our initial disconnection proposed above becomes the second stage of our retrosynthetic analysis. The first step is in fact functional group interconversion (FGI) of not only the secondary alcohol to a ketone but also the primary alcohol to an ester (8) as shown in Figure 8.2.

When analysing a molecule using the retrosynthetic approach it is important to always bear in mind the forward synthesis and to be aware of the presence of other functional groups within the molecule. Therefore in this case it is advantageous to make use of the fact that not only will reduction of a ketone give a secondary alcohol but reduction of an ester would give a primary alcohol, i.e. the two functional groups in our target molecule. The precise utility of this transformation will become apparent later.

The next stage in the retrosynthetic analysis of (2) involves disconnection (labelled 2) of the carbon–nitrogen bond in (9), to give synthons (10) and (11) equivalent to ethyl acrylate and the secondary amine (12) respectively. Although retrosynthetic analysis is a valuable tool in planning a synthesis of a target molecule, it may not be applied successfully at all times. The synthesis of amine (12) is an ideal example of this! Adding the latent polarities to (12) starting from either the amine or the two ester functions leads to a dissonant pattern. Therefore it is not apparent which disconnection to make. However the amine (12) has been successfully prepared by Roder and co–workers as shown in Figure 8.3.[2]

Therefore in Figure 8.2 we have designed a plausible retrosynthetic analysis of platynecine (**2**) which obeys the rules outlined in this book. The synthesis of the target molecule is shown in Figure 8.3.

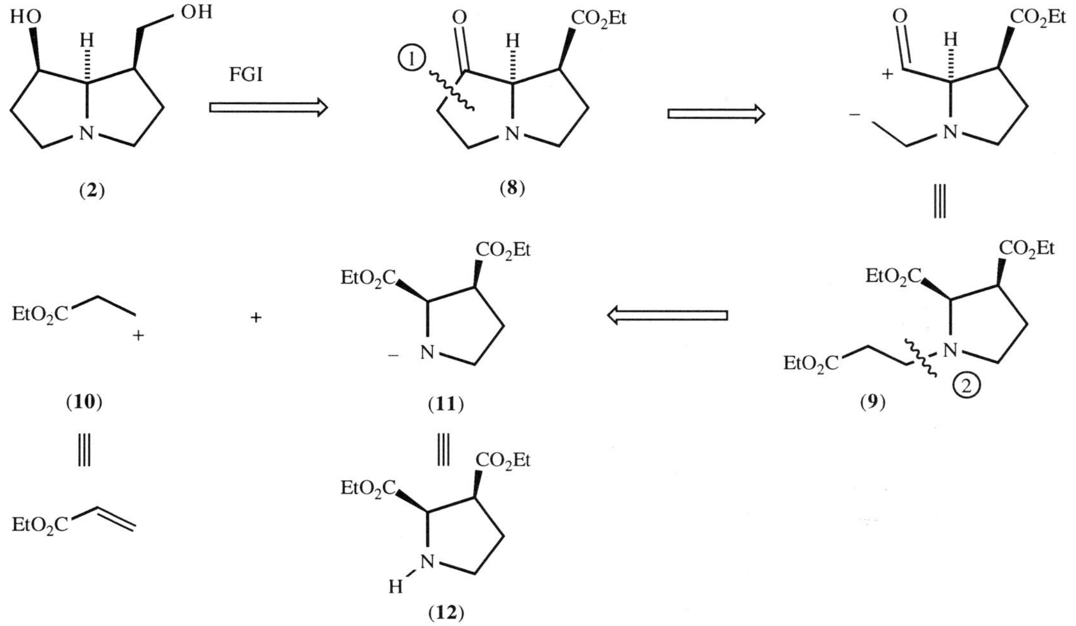

Figure 8.2 Retrosynthetic analysis of platynecine (**2**)

Synthesis

The first step in the synthesis of the target molecule (**2**) (Figure 8.3) was the nucleophilic displacement of the iodide in (**13**) by dibenzylamine to cleanly give (**14**). *(Why is it necessary to protect the amine?)* Reaction of (**14**) with ethyl oxalyl chloride in the presence of a base, potassium *t*–butoxide, gave the required keto–diester (**15**) in 82% yield. A series of reactions then took place in one–pot. Firstly the dibenzyl protecting group in (**15**) was reductively cleaved by catalytic hydrogenolysis to give the free amine which then reacted with the ketone to give a cyclic imine. Rearrangement of the imine gave (**16**) which was reduced to the required saturated diester (**12**).

Conjugate addition of amine (**12**) to ethyl acrylate gave (**17**) which cleanly undergoes a Dieckmann cyclisation in the presence of sodium hydride to give the required bicyclic framework of platynecine with the correct stereochemistry for the ring junction and the ester at C–1.

Decarboxylation of the β–keto ester function in (**18**) was achieved with potassium hydroxide at reflux and the resulting ketone was not isolated but immediately reduced with sodium borohydride to give the hydroxy acid which lactonised spontaneously to give (**19**). Sodium borohydride approaches from the less hindered α–face of the ketone giving the required β–alcohol which lactonised. Finally lithium aluminium hydride reduction of the lactone (**19**) gave the target molecule platynecine (**2**) in eight steps and 13% overall yield.

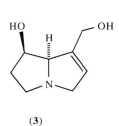

Figure 8.3 Synthesis[1,2] of platynecine (2)

8.4 Retronecine (3)–method 1

Retrosynthetic analysis

Unlike the retrosynthetic analysis of platynecine (2) described above, this analysis begins by considering a disconnection of type B as described in Section 8.2. Two functional group interconversions need to be considered prior to this disconnection:

i. Conversion of the primary alcohol to the ester (the forward step being reduction).

ii. Conversion of the double bond in (3) to the secondary alcohol which bears a consonant relationship with the ester. The equivalent forward synthetic step is dehydration of this secondary alcohol to the required olefin.

These two FGIs lead to the dihydroxy–ester (20).

Disconnection 1, the carbon–carbon bond in (20) (Figure 8.4), gives the synthon (21) in which a negative charge resides α to the ester (simple abstraction of the acidic proton) and the positive charge resides on the carbon of the carbon–oxygen bond, giving (22) as the synthetic equivalent to synthon (**21**). However due to the availability of starting materials, it proved simpler to make the synthetic equivalent to the positive charge in (**21**) an ester rather than an aldehyde, i.e. (**23**) rather than (**22**) (Figure 8.5).

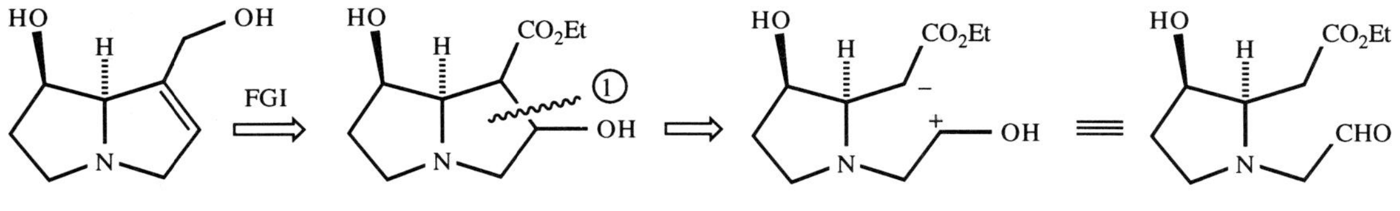

Figure 8.4 Retrosynthetic analysis of (**3**)

As we saw in the formation of (**19**) (Figure 8.3), a facile lactonisation reaction may occur between an alcohol and an ester if they are positioned in a molecule such that the intramolecular reaction leads to the formation of a 5–membered ring lactone (known as a γ–lactone). The reverse reaction is also possible, i.e. treatment of a lactone with sodium ethoxide returns the hydroxy–ester. Since intramolecular cyclisation of an alcohol and ester to give a γ–lactone is such a facile process (and indeed may be difficult to avoid when the molecule is subjected to basic reaction conditions), the next stage of the retrosynthetic analysis of retronecine (**3**) involves a FGI to convert the hydroxy diester (**23**) to a lactone (**24**) (Figure 8.5). Disconnection 2 in (**24**) then gives synthons (**25**) and (**26**), the synthetic equivalents of which are the secondary amine (**27**) and commercially available ethyl 2–bromoethanoate (**28**) respectively. FGI of the lactone (**27**) via cleavage of the 5–membered ring lactone back to the hydroxy–ester followed by conversion of the resultant secondary alcohol to a ketone, gives the keto–ester (**29**).

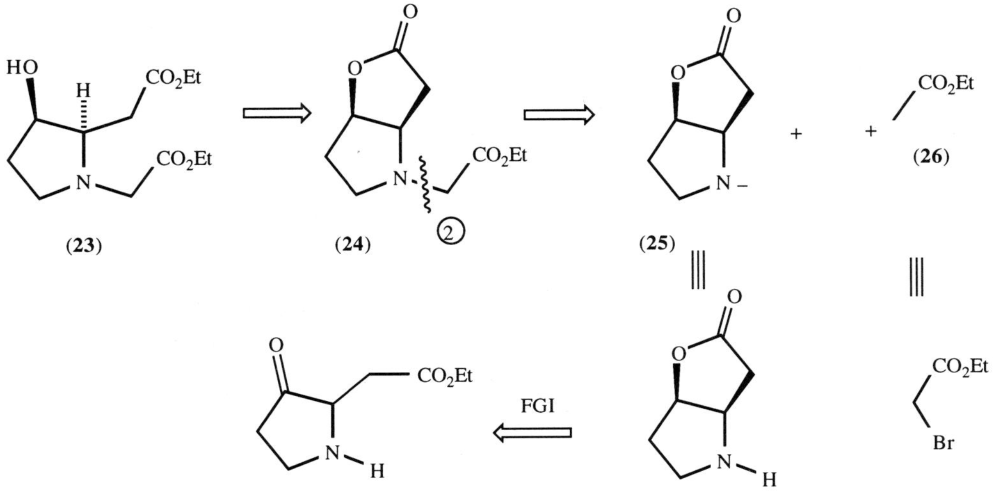

Figure 8.5 Retrosynthetic analysis of (**3**) continued

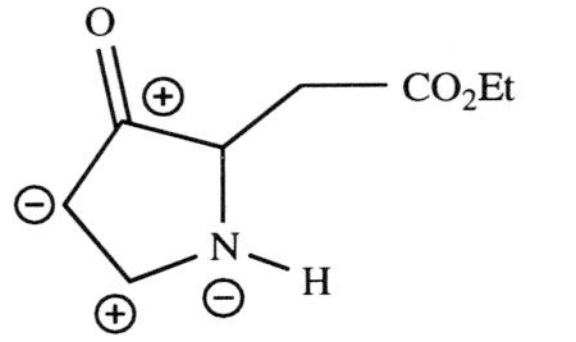

(29)

Unlike the retrosynthetic analysis of amine (12) in the previous synthesis (Figure 8.3) which did not aid the design of a synthetic route because of a dissonant pattern of latent polarities, adding the latent polarities to (29) leads to a consonant pattern between the ketone and the nitrogen. Working backwards, first it is necessary to protect the amine function (see below) and then, in theory, two disconnections could be considered in (34), either the C–N bond (labelled A) or the C–CO bond (labelled B) (Figure 8.6). The latter was selected. In order to obtain a synthetic equivalent with a negative charge at the required position, an ester (32) (Figure 8.7) was used. The use of this ester is analogous to the use of the ester (7) for the synthesis of platynecine described in the previous section.

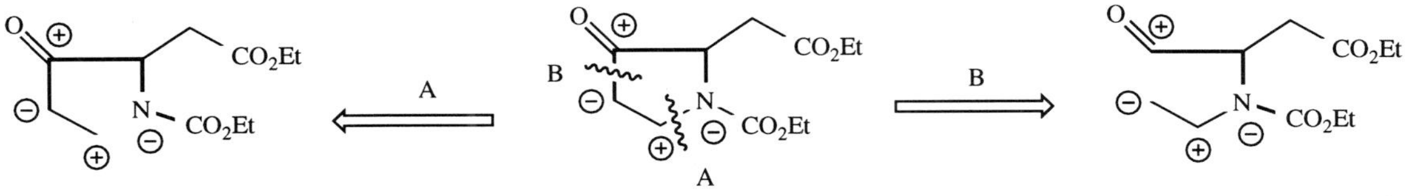

(34)

Figure 8.6 Retrosynthetic analysis of (34)

The retrosynthetic analysis outlined above is a combination of two approaches to the synthesis of retronecine. T.A. Geissman and A.C. Waiss[3] developed the route as far as lactone (24) and this was then used by K. Narasaka and his co–workers[4] to prepare the target molecule (3).

Synthesis

As with the synthesis of platynecine described above, it is necessary to initially protect the amine to prevent the formation of unwanted by–products in the first reaction. In this case, the authors favoured the use of the N–ethoxycarbonyl derivative from which the free amine may be unmasked with barium hydroxide.

The synthesis of (3) is shown in Figure 8.7. Conjugate addition of the protected amine (30) to the unsaturated diester (31) (diethyl fumarate) gave the triester (32) which spontaneously cyclised via a Dieckmann reaction to give (33). Hydrolysis of (33) followed by decarboxylation of the resultant β–keto acid and re–esterification of the remaining acid function gave the ketone (34) in 45% overall yield.

Chemoselective reduction of the ketone in (34) with sodium borohydride gave a secondary alcohol which spontaneously cyclised to give the lactone (35). Deprotection of the amine with barium hydroxide proceeded smoothly to give (27) in 70% yield. The next stage of the synthesis involved alkylation of the amine (27) with ethyl bromoacetate giving the N–carboethoxymethyl lactone (24) which was used by Narasaka and his co–workers to complete the synthesis of the target molecule, retronecine (3).

Treatment of the bicyclic lactone (24) with sodium ethoxide gave a hydroxy ester which on reaction with sodium hydride gave the keto ester (36) via a Dieckmann cyclisation. The keto ester (36) was not isolated but reduced *in situ* to the dihydroxy ester (37).

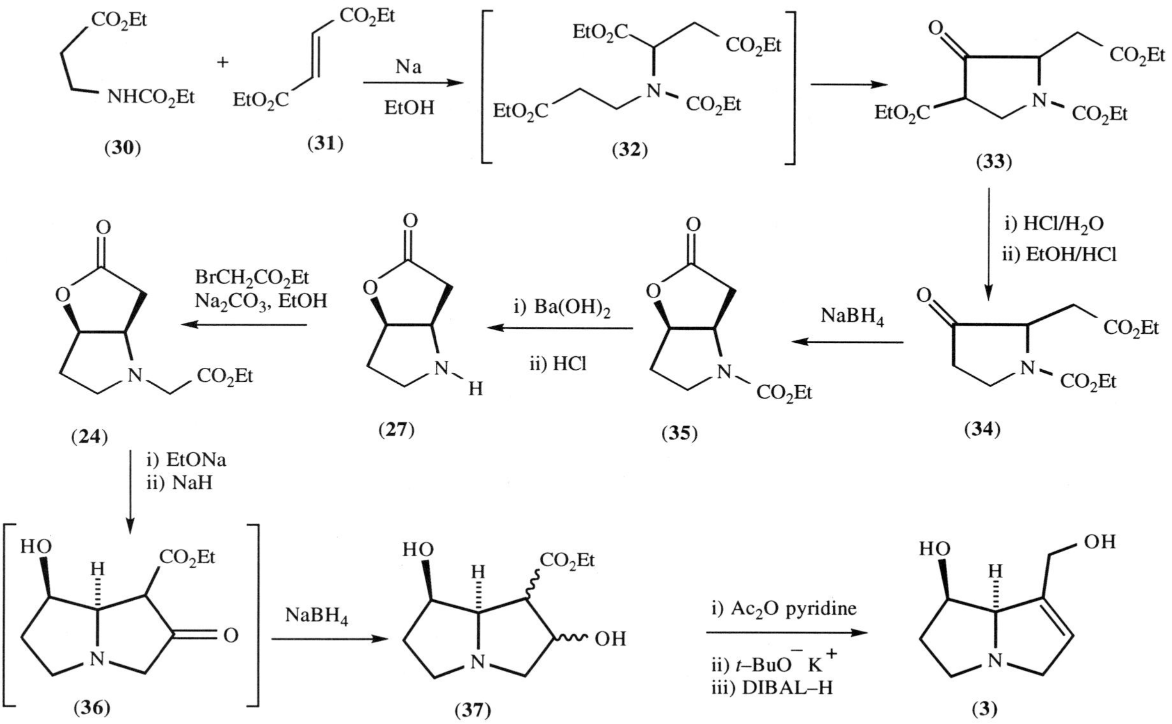

Figure 8.7 Synthesis of retronecine (3)

The final stages of the synthesis of retronecine (3) require the formation of the double bond and reduction of the ester function to the corresponding primary alcohol. Initially the diol (37) was converted to a diacetate by reaction with acetic anhydride in pyridine. Treatment of the diacetate with potassium *t*–butoxide gave elimination of the acetate β to the ester to generate the α,β–unsaturated ester in 71% yield. *Why is this acetate selectively eliminated whilst the other secondary acetate remains unaffected?* Reduction of the ethyl ester and the acetate with DIBAL–H gave the target compound (3) in 2% overall yield.

8.5 Retronecine (3)–method 2

Retrosynthetic analysis

The next synthesis to be considered is again of retronecine (3), but in this case we will use disconnection C (section 8.2), the cleavage of the carbon–nitrogen bond in the azabicyclo[3.3.0]octane framework, as the first disconnection.

During the last twenty years several methods have been developed for the formation of carbon–nitrogen bonds including S_N2 attack of nitrogen to a carbon atom bearing a suitable leaving group. In addition in this chapter, we have illustrated the value of the condensation of an amine with a ketone to give an imine which in turn may be reduced to give the C–N bond (Figure 8.3). In the approach adopted in the synthesis described below, cyclisation to form the basic necine skeleton is achieved using a C–N bond forming reaction involving electrophilic attack of benzenesulphenyl chloride to an olefin to give the epi–sulphonium ion which in turn is attacked by nitrogen (Figure 8.8).

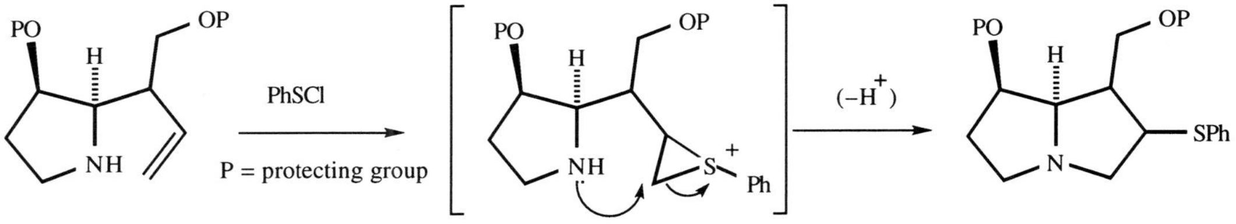

Figure 8.8

The first stage of the retrosynthetic analysis (Figure 8.9) of retronecine (**3**) is a FGI involving protection of the two hydroxyl functions and addition of a phenyl thioether across the double bond to generate (**38**). *Why is it necessary to protect the hydroxyl groups?* Disconnection of the C–N bond in (**38**) gives synthon (**39**) which, as we have seen above, may be considered as equivalent to the unsaturated amine (**40**).

Following a FGI of the protected secondary alcohol in (**40**) to a ketone, disconnection 2 gives two synthons (**41**) and (**42**). It is necessary to use this FGI in order to obtain the synthon (**41**) in which a negative charge may be generated α to nitrogen. The synthon (**42**) bearing a positive charge is equivalent to (**43**) (Figure 8.10) which is a known compound simply prepared by the method of Danishefsky and Regan.[5] The mechanism of this coupling reaction is discussed below.

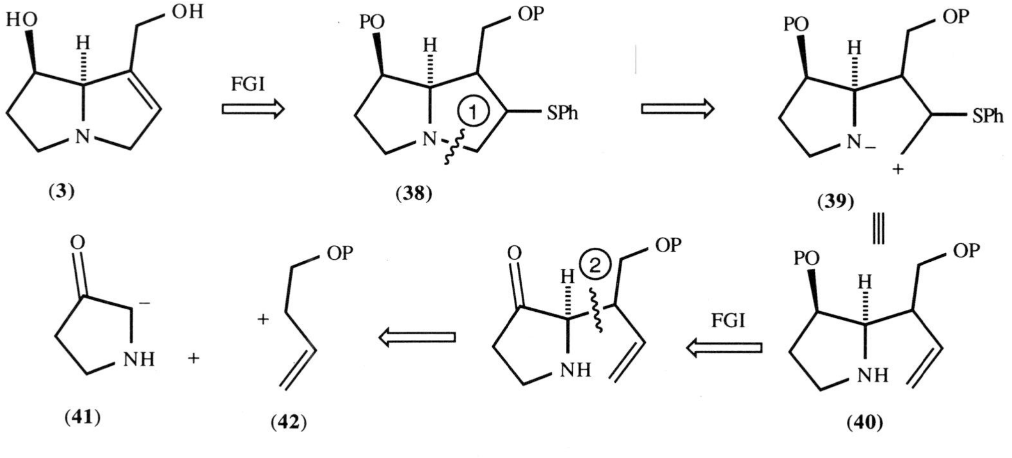

Figure 8.9 Retrosynthetic analysis 2 of retronecine (**3**)

Synthesis

The synthesis of retronecine based on this analysis was achieved by Ohsawa and his co–workers[6] as follows (Figure 8.10). The 3–pyrrolidone (**44**) was simply prepared from the diester (**45**) in 59% yield by firstly protecting the amine with ethyl chloroformate followed by a base induced Dieckmann cyclisation and finally hydrolysis and decarboxylation of the resultant β–keto acid.

The 3-pyrrolidone (**44**) was then treated with the protected *(Z)*–1,4–dihydroxybut–2–ene (**43**) in the presence of an acid catalyst to give the unsaturated ketone (**46**) in 77% yield. The reaction is believed to proceed by a [3,3] sigmatropic rearrangement of the allyl enol ether intermediate (**47**) via a chair–like transition state with total regio– and stereo– control.

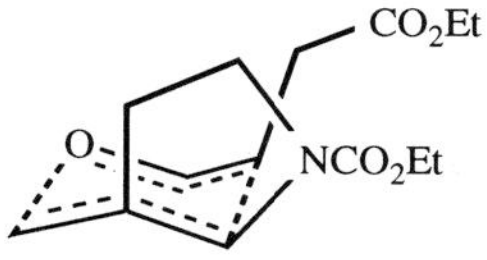

Reduction of the ketone (**46**) with sodium borohydride occurs predominantly from the less hindered face to give the required β–alcohol which was immediately protected as the benzyl ether (**48**). The next stage of the synthesis was the sulphenylamination reaction outlined in Figure 8.8. This was achieved by initial hydrolysis of the carbamate group in (**48**) with potassium hydroxide to give the secondary amine which was then converted to its hydrochloride salt prior to treatment with benzenesulphonyl chloride. The resultant ion was treated with base to effect the cyclisation reaction giving the required phenyl thioether (**49**) (equivalent to (**38**) in the retrosynthetic scheme Figure 8.9) as a single isomer in 72% yield.

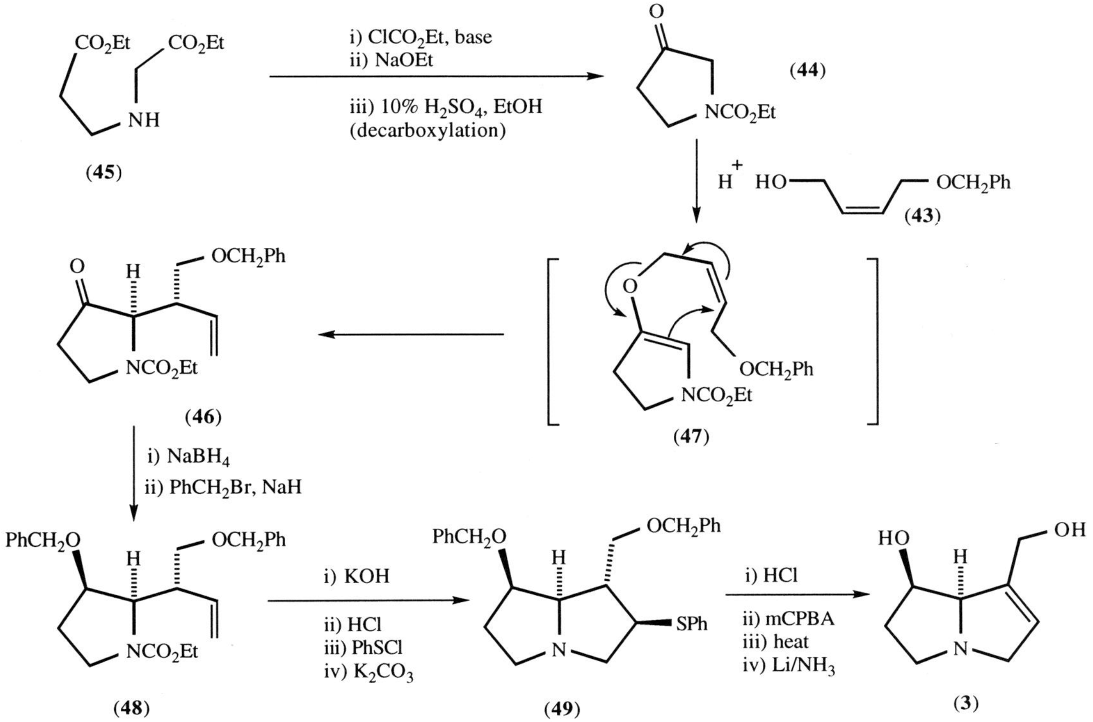

Transition state of (**47**)

Figure 8.10 Synthesis 2 of retronecine (**3**)

For the next stage of the reaction sequence it is crucial that the observed diastereomer (**49**) is formed in this cyclisation step (see below). *Why is diastereomer (49) formed?*

The final stages of the synthesis of retronecine involved elimination of the thioether function and deprotection of the primary and secondary alcohols. It is well established that oxidation of a thioether to a sulphoxide followed by thermolysis results in a *syn* elimination process to give an alkene. Thus treatment of the phenyl thioether (**49**) with the oxidising agent *m*–chloroperbenzoic acid and heating the resultant sulpoxide in refluxing xylene gave the protected retronecine (**50**) (Figure 8.11).

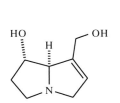

Figure 8.11 Elimination of the sulphoxide

Finally deprotection of the alcohols by reductive cleavage of the benzyl ethers with lithium in liquid ammonia gave the target molecule (**3**).

8.6 (+)-Heliotridine (4)

Retrosynthetic analysis

The final synthesis to be considered in this chapter involves the enantioselective (or asymmetric) synthesis of another dihydroxylated pyrrolizidine alkaloid, (+)–heliotridine (**4**). As mentioned in the introduction, enantioselective synthesis is now of great importance to the pharmaceutical and agrochemical industries. Although products may eventually be sold as racemates it is obviously important to be aware of the properties of both enantiomers in the mixture before a particular drug is prescribed.

There are a number of methods to effect the synthesis of a homochiral molecule, for example the use of:

i) Homochiral starting materials
ii) Homochiral reagents/auxiliaries
iii) Enzymatic methods.

The synthesis of (+)–heliotridine described below uses a homochiral starting material, (*S*)–malic acid. The approach to (+)–heliotridine involves the disconnection D (introduced in section 8.2) of the azabicyclo[3.3.0]octane skeleton.

Heliotridine (**4**)

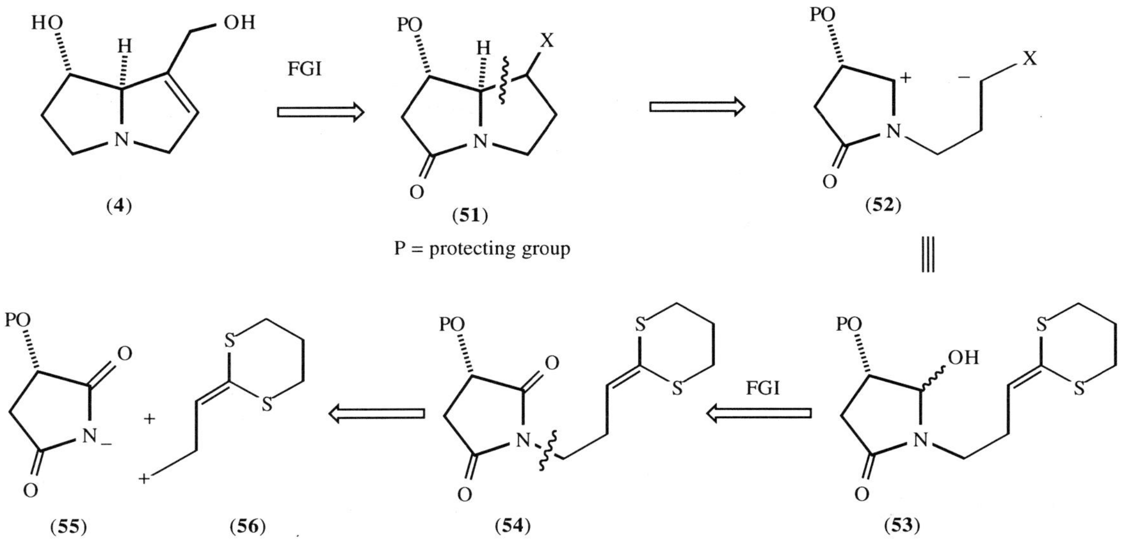

Figure 8.12 Retrosynthetic analysis of (+)–heliotridine (4)

In common with the other three retrosyntheses that we have now considered, the first stage of the analysis of (4) (Figure 8.12) involves a FGI. In this case the secondary alcohol must be protected and in place of the primary alcohol, a functional group (X) is required to give the synthetic equivalent to the negative charge of the disconnected carbon–carbon bond highlighted in (51). Several functional groups could have been considered to serve this purpose e.g. X may be an ester. However the dithiane was favoured by Chamberlin and Chung.[7] The positive charge of synthon (52) requires a good leaving group α to nitrogen, and a hydroxyl group was used. Hence the synthetic equivalent of synthon (52) is (53). FGI on (53) gives the imide (54). The necessity of this FGI is not immediately apparent but relates to the availablity of starting materials.

Disconnection of the carbon–nitrogen bond of (54) gives two fragments (55), with a negative charge on nitrogen and the positively charged synthon (56). Due to the availability of starting materials, the most easily accessible homochiral synthetic equivalent of (55) proved to be the succinimide (57) (Figure 8.13) which may, in turn, be prepared from commercially available (S)-malic acid and ammonia as described below. A synthetic equivalent to (56) is 2–(3–hydroxypropylidine)–1,3–dithiane (58) which may be prepared in good yield by the method of Chamberlin and Chung.[7]

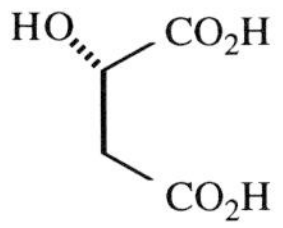

Synthesis

In the first stage of the synthesis of (+)–heliotridine (Figure 8.13) (S)–malic acid was treated sequentially with acetyl chloride, gaseous ammonia and then acetyl chloride again to give the (S)-acetoxysuccinimide (57) in 52% yield after crystallisation.

(S)–Malic acid

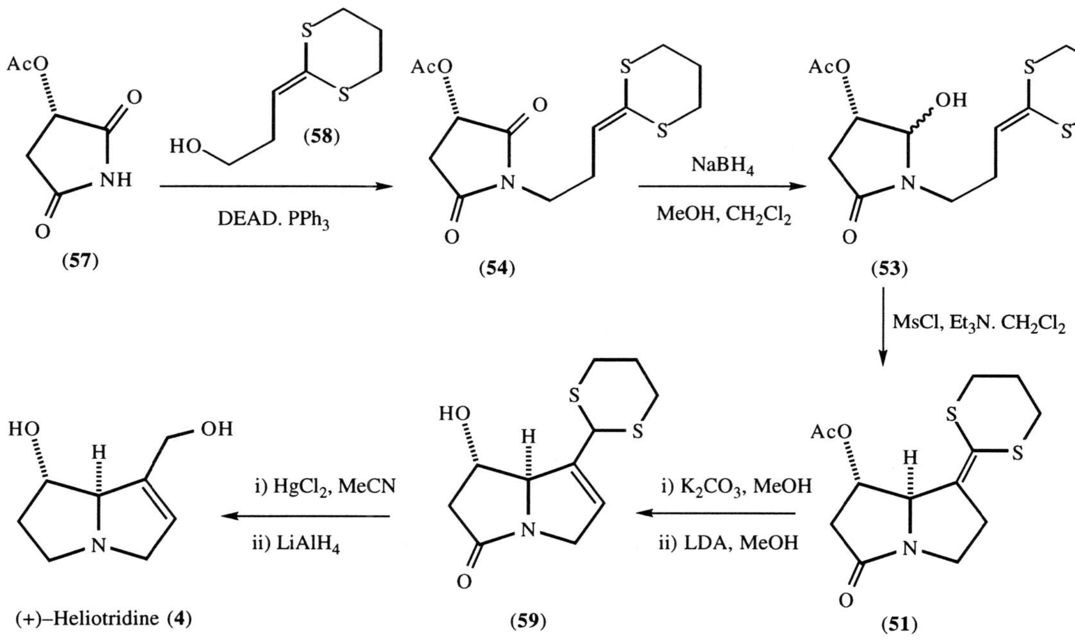

Figure 8.13 Synthesis of (+)–heliotridine (4)

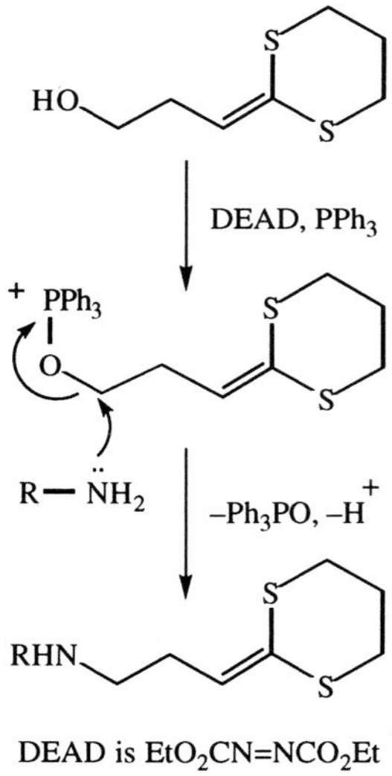

DEAD is EtO$_2$CN=NCO$_2$Et

Figure 8.14

The imide (57) was coupled with 2–(3–hydroxypropylidene)–1,3–dithiane in good yield. It was necessary to activate the alcohol in this reaction and Chamberlin and Chung[7] used a method first described by Mitsunobu which is outlined in Figure 8.14.

Reduction of the diimide (54) with sodium borohydride gave (53) in 85% yield. This reaction was carefully monitored since succinimides are sensitive to over–reduction by sodium borohydride.

The key ring forming step was then achieved in 68% yield by firstly converting the alcohol to a good leaving group (a mesylate) which then spontaneously cyclised to (51) by the mechanism illustrated in Figure 8.15.

Having formed the pyrrolizidine ring system, the next operation was to hydrolyse the acetate to free the secondary alcohol using potassium carbonate and then to migrate the double bond to the endocyclic position via a deprotonation–protonation reaction with LDA. These two steps were achieved to give (59) in 51% yield.

Finally hydrolysis of the dithiane to the aldehyde and reduction of both the aldehyde and the lactam carbonyl groups with lithium aluminium hydride gave (+)–heliotridine (4) in 8% overall yield from commercially available (*S*)–malic acid.

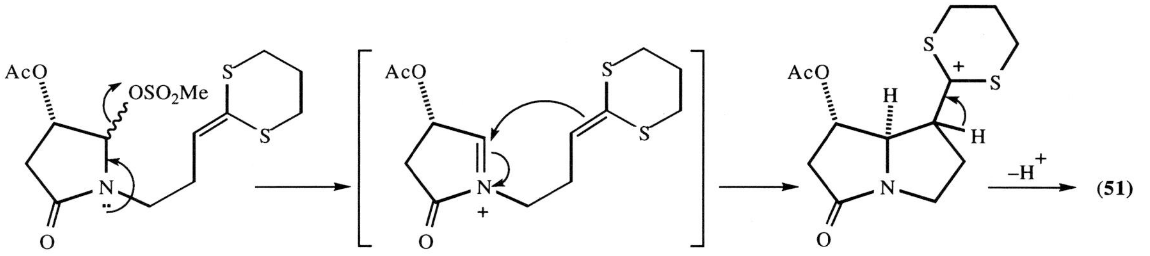

Figure 8.15 Ketene thioacetal cyclisation

Concluding remarks

Four methods for the synthesis of the pyrrolizidine alkaloids have been described using four different initial disconnections of the azabicyclo[3.3.0]octane framework. It would be difficult to select which is the best method. Each route has its own particular merits, e.g. the route to platynecine gives the highest yield, the second method for the synthesis of retronecine uses stereoselectivity to good effect and of course, the route to (+)–heliotridine is the only enantioselective synthesis. As stated on page one of this primer: 'When considering possible routes for the synthesis of a particular target molecule, one draws on an almost unimaginably vast database of potential transformations and for this reason the subject remains largely a matter of personal opinion and interpretation'.

References

1. E. Roder, T. Bourauel and H. Wiedenfeld, *Liebigs Ann. Chem.*, 1990, 607.
2. E. Roder, H. Wiedenfeld and T. Bourauel, *Liebigs Ann. Chem.*, 1987, 1117.
3. T. A., Geissman and A. C. Waiss Jr., *J. Org. Chem.*, 1962, **27**, 139.
4. K. Narasaka, T. Sakahura, T. Uchimaru and D. Guedin-Vuong, *J. Am. Chem. Soc.*, 1984, **106**, 2954.
5. S. Danishefsky and J. Regan, *Tetrahedron Letters*, 1981, **22**, 3919.
6. T. Ohsawa, M. Ihara and K. Fukumoto, *J. Org. Chem.*, 1983, **48**, 3644.
7. A.R. Chamberlin and J. Y. L. Chung *Tetrahedron Letters*, 1982, **23**, 2619; *J. Am. Chem. Soc*, 1983, **105**, 3655; *J. Org. Chem*, 1985, **50**, 4425.

Glossary of Terms

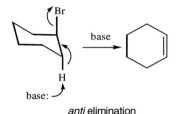

anti elimination

Absolute configuration The specification of *R* or *S* at each asymmetric centre in a molecule.

1,2–Addition Addition of a nucleophile to the carbonyl carbon of an α,β–unsaturated carbonyl compound.

anti (i.e. opposed) A term similar in meaning to *trans*, but used particularly in naming certain classes of compounds and reactions e.g. *anti* elimination

Asymmetric (or stereogenic) centre A tetrahedral atom with four different groups attached to it. In the case of third row elements a nonbonding pair of electrons may count as one of the groups.

Chemoselective reaction A process in which one functional group within a molecule reacts leaving further potentially reactive functionality unaltered.

cis (Latin, on the same side). A term similar in meaning to *Z*, but referring to identical atoms or groups on the same side of a double bond. Also used to describe substituents positioned on the same face of a ring.

Configuration The fixed relative spatial arrangement of atoms in a molecule.

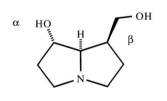

α– and β– configuration The α–configuration is used to denote the downward substituent position and the β–configuration to denote the upward substituent position in rigid molecules such as the pyrollizidine alkaloids. *n.b.* It may be a source of some confusion that α, β, γ etc are also used to denote substituent positions with respect to functional groups.

Conformation A particular orientation of the atoms in a molecule, differing from other possible orientations by rotation(s) around single bonds. Unless it is held rigid by small rings or double bonds, a molecule may have an infinite number of conformations, but only one configuration.

Conjugate (or 1,4–) addition   Addition of a nucleophile to the β–position of an α,β–unsaturated carbonyl compound. If the addition involves attack by a carbon nucleophile, this is known as Michael addition or the Michael reaction.

Diastereomers/diastereoisomers Stereoisomers that are non–enantiomers. A compound with n stereocentres can have a maximum of 2^n diastereoisomers.

Disconnection   The imagined cleavage of a bond to 'break' the molecule into possible starting materials during retrosynthetic analysis.

E- alkene

E (entgegen) (German–opposite) The arrangement in which the two higher priority groups (and the two lower priority groups) assigned by the Cahn–Ingold–Prelog system are on opposite sides of a double bond.

Enantiomers   Two molecules that are related as object and nonsuperimposable mirror image.

Enantioselective reaction   A reaction is which one enantiomer is produced (or destroyed) more rapidly than the other; the amount by which one enantiomer exceeds the other in a mixture is known as the enantiomeric excess (ee).

Erythrose

Epimers Diastereomers differing in configuration at only one (of several) asymmetric centre.

erythro   A prefix given to the name of a diastereomer with configuration similar to the two adjacent asymmetric centres of the four carbon aldose erythrose, i.e. *S, S* or *R, R*.

Endo-7-methyl-2-norcamphor

endo That postion in a bicyclic molecule opposite the main bridge. The *endo* substituent is generally considered to be in the 'inside' or more hindered position. The main bridge is determined by applying the priorities (highest first): i) bridge with heteroatoms; ii) bridge with fewest members; iii) saturated bridge; iv) bridge with fewest substituents; v) bridge with lowest priority substituents.

Exo-7-methyl-2-norcamphor

exo That position in a bicyclic molecule nearer the main bridge. The *exo* substituent is often considered to be on the 'outside' or less hindered position.

FGI Functional group interconversion is the process of converting one functional group into another during retrosynthetic analysis.

Inversion of configuration The conversion of one molecule into another that has the opposite relative configuration.

Latent polarity

I   II

Stereoisomers

Latent polarity The imaginary pattern of alternating positive and negative charges used to assist in the identification of suitable disconnections and synthons when using retrosynthetic analysis. Latent polarities are arranged such that a positive charge is placed on the carbon adjacent to the heteroatom.

R (***rectus***) That configuration of an asymmetric centre with a clockwise relationship of group priorities (highest to second lowest) using the Cahn–Ingold–Prelog system, when viewed along the bond towards the lowest priority group.

re* and *si In trigonal molecules containing for example a carbonyl group or an alkene, the two faces can be named by an extension of the Cahn–Ingold–Prelog system. If the three groups are arranged by the sequence rules that have the order X > Y > Z, that face in which the groups in the sequence are clockwise (as in I) is the *re* face (from Latin–*rectus*) whereas II shows the *si* face (from Latin–*sinster*).

Regioselective reaction A reaction in which the formation of one structural (or positional) isomer is favoured over another in a mixture.

Retention of configuration The conversion of one molecule into another with the same relative configuration; the opposite of inversion of configuration.

Retrosynthetic analysis The process of breaking down a target molecule (TM) into available starting materials by disconnections and/or functional group interconversions (FGI).

S (***sinister***) That configuration of an asymmetric centre with an anticlockwise relationship of group priorities (highest to second lowest) when viewed along the bond towards the lowest priority group.

Stereochemistry A description of the three dimensional spatial relationships between atoms in a molecule.

Stereoisomers Molecules having the same sequence of atoms and bonds, but differing in the fixed three dimensional arrangement of these atoms. Each stereoisomer has a unique configuration that can only be converted to a different configuration by chemical means (i.e. making and breaking of bonds). There are two distinct classes of stereoisomers: enantiomers and diastereomers.

Stereoselective reaction A reaction in which one stereoisomer in a mixture is produced (or destroyed) more rapidly than another, resulting in a predominance of the favoured stereoisomer in the mixture of products.

Stereospecific reaction A process in which a particular stereoisomer reacts to give one specific stereoisomer of product, e.g. in an S_N2 reaction.

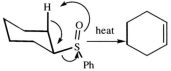

Syn elimination

syn (Greek –together) A term similar in meaning to Z or *cis*, used in naming certain classes of compounds and reactions, e.g. *syn* elimination

Synthetic equivalent A reagent carrying out the function of a synthon.

Synthon An idealised fragment (usually a cation or anion) resulting from a disconnection during retrosynthetic analysis.

Target molecule The compound to be synthesised.

threo A prefix given to the name of a diastereomer with configuration similar to the two adjacent asymmetric centres of the four carbon aldose threose, i.e. *R, S* or *S, R*.

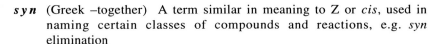

Threose

trans (Latin, across). A term similar in meaning to E, but referring to identical atoms or groups on opposite sides of a double bond. Also used to describe substituents positioned on opposite sides of a ring.

Umpolung The German word umpolung is used to describe cases in which a synthon of opposite polarity to that normally associated with a required functional group must be used.

Z (zusammen) (German–together) The arrangement in which the two higher priority groups (and two lower priority groups) are on the same side of a double bond.

Z- alkene

Index